CONTENTS

Make: **Volume 67** Feb/March 2019

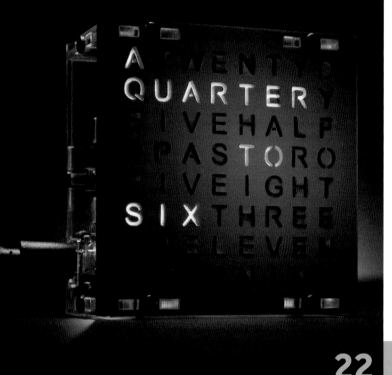

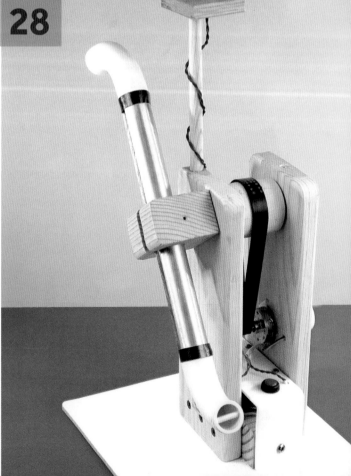

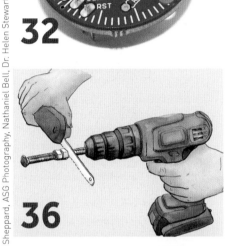

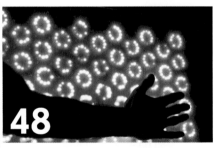

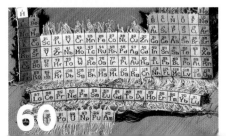

Mark Madeo, Jordi Bover, Drew Nikonowicz, Charles Platt, Petr Maur, Adafruit, Richard Sheppard, ASG Photography, Nathaniel Bell, Dr. Helen Stewart

ON THE COVER:
For its debut last November, La Machine's Minotaur took part in an elaborate play over the course of several days throughout the streets of Toulouse, France.

Photo: Jordi Bover

Make:

WELCOME

EXECUTIVE
CHAIRMAN & CEO
Dale Dougherty
dale@makermedia.com

CFO & COO
Todd Sotkiewicz
todd@makermedia.com

EDITORIAL

EDITORIAL DIRECTOR
Roger Stewart
roger@makermedia.com

EXECUTIVE EDITOR
Mike Senese
mike@makermedia.com

SENIOR EDITORS
Keith Hammond
khammond@makermedia.com
Caleb Kraft
caleb@makermedia.com

EDITOR
Laurie Barton

PRODUCTION MANAGER
Craig Couden

BOOKS EDITOR
Patrick Di Justo

CONTRIBUTING EDITORS
William Gurstelle
Charles Platt
Matt Stultz

CONTRIBUTING WRITERS
Julia Jameson Barton, Nathaniel Bell, Gareth Branwyn, Chris Connors, Tyler Cooper, Andy Doro, Jordan Ficklin, Limor Fried, Jake Guttormsson, Sam Guyer, Mikaela Holmes, Bob Knetzger, Darrell Maloney, Drew Nikonowicz, Becky Stern, Eric Steuer, Jane Stewart, Morten Nisker Toppenberg

DESIGN, PHOTOGRAPHY & VIDEO

ART DIRECTOR
Juliann Brown

SENIOR VIDEO PRODUCER
Tyler Winegarner

MAKEZINE.COM

WEB/PRODUCT DEVELOPMENT
Rio Roth-Barreiro
Maya Gorton
Pravisti Shrestha
Stephanie Stokes
Travis Stone
Alicia Williams

CONTRIBUTING ARTISTS
Jordi Bover, Bob Knetzger, Mark Madeo

ONLINE CONTRIBUTORS
Jennifer Blakeslee, Becky Button, Briana Campbell, Jon Christian, David Cole, DC Denison, Gretchen Giles, Bob Goldstein, Kathryn Jezer-Morton, Grete Kaulinyte, Mark Mathias, Scott N. Miller, Goli Mohammadi

PARTNERSHIPS & ADVERTISING
makermedia.com/contact-sales or partnerships@makezine.com

SENIOR DIRECTOR OF PARTNERSHIPS & PROGRAMS
Katie D. Kunde

DIRECTOR OF PARTNERSHIPS
Shaun Beall

STRATEGIC PARTNERSHIPS
Cecily Benzon
Brigitte Mullin

DIRECTOR OF MEDIA OPERATIONS
Mara Lincoln

DIGITAL PRODUCT STRATEGY

SENIOR DIRECTOR, CONSUMER EXPERIENCE
Clair Whitmer

MAKER FAIRE

MANAGING DIRECTOR
Sabrina Merlo

WEB PRODUCER
Bill Olson

MAKER SHARE

DIGITAL COMMUNITY PRODUCT MANAGER
Matthew A. Dalton

COMMERCE

PRODUCT MARKETING MANAGER
Ian Wang

OPERATIONS MANAGER
Rob Bullington

PUBLISHED BY

MAKER MEDIA, INC.
Dale Dougherty

Copyright © 2019
Maker Media, Inc. All rights reserved. Reproduction without permission is prohibited.
Printed in the USA by Schumann Printers, Inc.

Comments may be sent to:
editor@makezine.com

Visit us online:
makezine.com

Follow us:
🐦 @make @makerfaire @makershed
google.com/+make
makemagazine
makemagazine
makemagazine
twitch.tv/make
makemagazine

Manage your account online, including change of address:
makezine.com/account
866-289-8847 toll-free in U.S. and Canada
818-487-2037,
5 a.m.–5 p.m., PST
cs@readerservices.makezine.com

Heartbreak Strikes Close to Home

BY MIKE SENESE, executive editor of *Make:* magazine

In the middle of our cycle for putting together this issue of *Make:*, a wildfire ignited a little over 100 miles from our offices. Unfortunately, it wasn't a quick one — over the ensuing days, the Camp Fire, as it would become known, turned into the most devastating blaze in California history, and the worst in the United States in over 100 years.

Choked by accumulating smoke and terrified by the videos and stories from inside the inferno, we, along with most of the region, struggled to focus on work as we hoped for the best for those affected by the flames.

Sadly, over 80 lives were lost and 18,000 buildings destroyed. Among those were homes, schools, and businesses, including a fledgling spot called A Maker's Space that provided creative education opportunities to the local community. *Make:* founder Dale Dougherty reached out to the proprietor Pamela Teeter to hear about the space, her family's harrowing story of escape, and what the future may hold for their endeavors; you can read Dale's article on our site: makezine.com/2018/11/20/a-makers-space-in-paradise-burns-down.

Our hearts go out to all those affected by this tragic incident, and other recent natural disasters. We're as alarmed as everyone about the increasing frequency and magnitude of these occurrences.

Are you or any makers you know involved in providing disaster relief solutions? I'd love to hear about them. Email me: mike@makermedia.com

CONTRIBUTORS

What's your favorite time travel machine and why?

Julia Jameson Barton
Versailles, MO
(Fantastic Plastic)
My favorite Time Machine is from the 2002 movie, *The Time Machine*. It was the design of the machine, with its Fresnel lenses glistening with magic that made me believe.

Andy Doro
Brooklyn, NY
(Time Will Tell)
Terry Gilliam's time travel machine in *12 Monkeys* is clunky, all tubes and plastic, as they send a criminal back in time in an attempt to save the world.

Mikaela Holmes
Las Vegas, NV
(Convertible Caddy)
Probably the Heart of Gold spaceship in *The Hitchhiker's Guide to the Galaxy*. The surreal physics of its Infinite Improbability Drive are just such a clever parody of the logic of spacetime travel.

Craig Couden
San Pablo, CA
(Toolbox)
Some might say the TARDIS from *Doctor Who*, but I'm going with Captain Jack Harkness' wrist mounted Vortex Manipulator. It's more portable, easy to keep on your person at all times, and makes for a quicker exit from sticky situations.

Issue No. 67, Feb/Mar 2019. *Make:* (ISSN 1556-2336) is published bimonthly by Maker Media, Inc. in the months of January, March, May, July, September, and November. Maker Media is located at 1700 Montgomery Street, Suite 240, San Francisco, CA 94111. SUBSCRIPTIONS: Send all subscription requests to *Make:*, P.O. Box 17046, North Hollywood, CA 91615-9588 or subscribe online at makezine.com/offer or via phone at (866) 289-8847 (U.S. and Canada); all other countries call (818) 487-2037. Subscriptions are available for $34.99 for 1 year (6 issues) in the United States; in Canada: $39.99 USD; all other countries: $50.09 USD. Periodicals Postage Paid at San Francisco, CA, and at additional mailing offices. POSTMASTER: Send address changes to *Make:*, P.O. Box 17046, North Hollywood, CA 91615-9588. Canada Post Publications Mail Agreement Number 41129568. CANADA POSTMASTER: Send address changes to: Maker Media, PO Box 456, Niagara Falls, ON L2E 6V2

PRINTED WITH SOY INK

Lofty Goals *and* Currency Concerns

We were so impressed with how Scorch
Works was able to levitate the disk of paper
["Easy Ultrasonic Levitation," *Make:* Vol. 65,
page 50] that we had to give it a try:
youtu.be/So2VK5xnLyI?t=3490.

You can also download files to 3D print
Scorch Works' custom Ultrasonic Levitator
Bracket at thingiverse.com/thing:3260376.

Ulrich Schmerold, Stacey Lee Webber, Courtesy of the Wazer Team

A PENNY FOR YOUR THOUGHTS

Sorry to bother but from the article ["Under
Pressure" *Make:* Vol. 66, page 16] there's
no disclaimer that points to the fact that
damaging U.S. currency is illegal.
moneyfactory.gov/resources/
lawsandregulations.html [states:]

Defacement of Currency

Defacement of currency is a violation of
Title 18, Section 333 of the United States
Code. Under this provision, currency
defacement is generally defined as follows:
Whoever mutilates, cuts, disfigures,
perforates, unites or cements together, or
does any other thing to any bank bill, draft,
note, or other evidence of debt issued by
any national banking association, Federal
Reserve Bank, or Federal Reserve System,
with intent to render such item(s) unfit to
be reissued, shall be fined under this title
or imprisoned not more than six months,
or both.

Visit secretservice.gov for additional
information.

—Jose Serrano, via email

Senior Editor Caleb Kraft Responds:
*Thanks for writing in, Joe! This is a topic that
we've encountered a few times over the years.
Art involving U.S. currency is actually legal as
long as the modifications aren't made with
fraudulent intent. This artist isn't changing
the coins in order to falsely convey a different
denomination, so they're fine.*

*You can read more on the treasury's website
here: treasury.gov/resource-center/faqs/
Coins/Pages/edu_faq_coins_portraits.aspx*

*Here's the specific section (emphasis added
by me):*

Is it illegal to damage or deface coins?

Section 331 of Title 18 of the United
States code provides criminal penalties
for anyone who "fraudulently alters,
defaces, mutilates impairs, diminishes,
falsifies, scales, or lightens any of the
coins coined at the Mints of the United
States." This statute means that you may
be violating the law if you change the
appearance of the coin and fraudulently
represent it to be other than the altered
coin that it is. As a matter of policy, the
U.S. Mint does not promote coloring,
plating or altering U.S. coinage: **however,
there are no sanctions against such
activity absent fraudulent intent**.

MADE ON EARTH

WELDED WHIMSY

TICKTOCKTOM.COM

It's not surprising that a guy who goes by the lighthearted name **Tick Tock Tom** would be into art primarily for the fun of it. "I'm not rich, but I'm amassing a really good collection of stories," says the Ottawa-bred sculptor. Many of those are about hunting down discarded metal and used machines, pulling apart the various pieces, and putting it all to use as source material for his artistic creations.

While the scrap metal sculptures Tom makes are highly intricate and truly beautiful, the dominant characteristics of his style are fun, humor, and wonder. Life-sized robots, creepy critters, wild masks … these are gorgeously cool art pieces you want to play with, not just look at in a gallery. The spirit with which Tom imbues his work stems from his childhood love of taking apart toys, home electronics, and computer equipment to see how the various parts inside all fit together — and how they might be combined to build something new.

Over the years, Tom has parlayed that early curiosity into a unique and bustling career making all manner of things derived from salvaged metal and other found materials. "Because of sculpture, I've been invited into so many other interesting types of projects," he says. In addition to his formal art practice, he's been commissioned to create work like props for TV and films; high-end wearable art pieces; and custom trophies for events like the Ottawa International Animation Festival.

One of the things Tom likes best about his work is the ability to turn lifeless materials like metal slabs and busted-up machinery into innovative creations that connect directly with the human heart. He recalls a moment at a recent show in Montreal, where a blind woman stopped in her tracks upon hearing the asthmatic wheezing of a set of mechanical lungs that Tom had on display. "I explained all about what she was listening to," he says. "She thanked me profusely, saying that since losing her vision, she could no longer appreciate art with her eyes."

—*Eric Steuer*

Petr Maur

INTRICATE INSECTS

INSTAGRAM.COM/HOMEMADETRAP

After a childhood visit to a zoological museum, **Emily Yeadon** of Yorkshire U.K., found herself mesmerized by the antique taxidermy, especially of the insects. The moths that she had seen therein were so vibrant and delicate, like little works of art. Over the years, Yeadon has taken that inspiration and refined and evolved it into these creations, which capture the beauty for the entomologically minded, without harming a single insect.

At first glance they seem alive, almost on the verge of flitting away into the night to do whatever it is that moths do, but alas they are actually made of fine wire and thread. Anatomically correct aside from scale, and visually stunning. Yeadon creates these completely by hand, bending and molding a wire skeleton, stitching wing patterns and, her favorite part, adding the minutiae; a dot of paint here or a bit of fuzz there.

—*Caleb Kraft*

Emily Yeadon

MEDIEVAL MACHINES WIMDELVOYE.BE

The renowned Belgian artist **Wim Delvoye** has a long history of bringing together the mundane and the ornate to create elaborate, often humorous art that encourages people to re-think how they see the world. In the late 1980s, he made a name for himself by applying the style and process used in Delftware (high-end blue and white Dutch pottery) to create beautifully decorative versions of everyday items like shovels and gas tanks. In 2000, he created *Marble Floors*, an installation featuring gorgeous and complex floor tiles — a close look at which would reveal that their designs were derived from scanned and printed reproductions of salami slices.

More recently, Delvoye has been expanding on an ongoing series entitled *Gothic Works*, which has been exhibited at many of the world's most famous and esteemed museums. For this project, the artist creates stained glass windows and detailed sculptures of modern objects in the style of International Gothic art. *Gothic Works* includes several miniature scale metal replicas of vehicles you'd find on a construction site — a dump truck, a cement truck, a flatbed trailer — which Delvoye makes by laser-cutting corten steel plates into intricate shapes characterized by slender piers and pointed arches. While Delvoye uses contemporary tools like modeling software to assist in designing the sculptures' latticework, the pieces are totally disorienting in their timelessness, and awe-inspiring in how elegantly they combine the cutting edge and the antiquated. —*Eric Steuer*

Studio Wim Delvoye, Belgium

STEELED FOR BATTLE

MICHAELJAYWILLIAMS.COM

Michael Jay Williams had been working with metal for decades before he discovered his talent for using it to create amazing art. During a personal rough patch, the New Jersey-based welder-fabricator became inspired to try his hand at using steel to express his creative side. "I was going through a divorce, and I needed something badly to channel my negative emotions into," he says. "So I started making my first metal sculpture." The experience was life changing for Williams, who has since created dozens of intricate sculptures, all by hand, by bending and shaping flat steel into freaky and fantastical forms.

His best-known and most breathtaking work is *A Mother's Love*, a massive sculpture that showcases Dottie the Fire-Breathing Dragon, an 11-foot-long metal beast that weighs in at nearly 600 pounds. Look closely and you'll see a baby dragon that's welded to the front leg, too. "Dottie's in defense mode,"

Williams used 375 feet of 0.25-inch solid round bar to create Dottie's body, then performed thousands of individual tack welds to give the dragon's "skin" surface a texture that feels scaly. He used 16-gauge carbon steel to build Dottie's wings, carefully heating each with a torch to add a colorful, faux-translucent finish. Finally, the artist built a propane-powered flame kit from black iron pipe to allow Dottie to bellow fire from her mouth.

Williams plotted the sculpture's form in his head and built the entire thing without ever sketching a plan on paper. In all, he spent 350 hours over a five-month period building Dottie, and did the bulk of the work in his personal studio, a small steel building not far from his home. "The end result might look like it required all kinds of tools," he says. "But all I needed was a grinder, a torch, a welder, and my hands." *—Eric Steuer*

David Casius · thedave.myportfolio.com

Give the Gift of Inspiration

2 FOR THE PRICE OF 1

Gift One, Get One
Make:

Go to makezine.com/subscribe
to get started

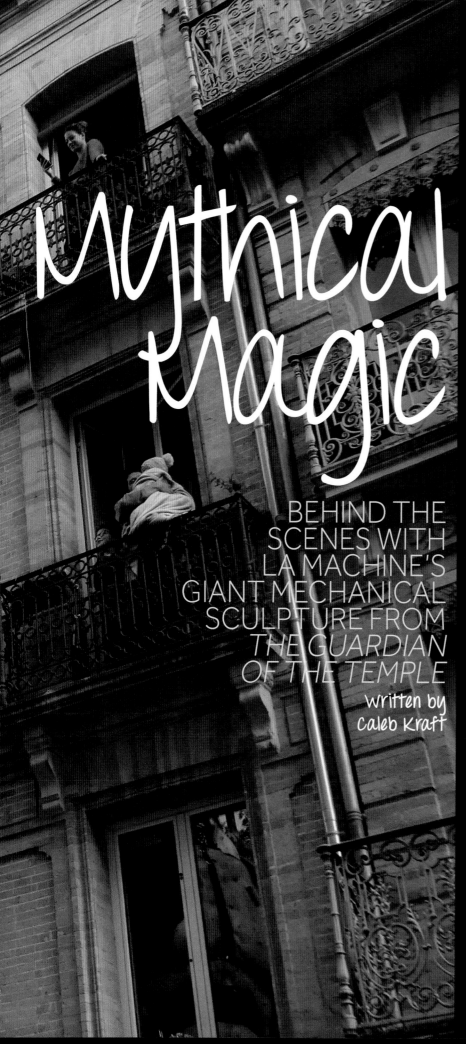

Mythical Magic

BEHIND THE SCENES WITH LA MACHINE'S GIANT MECHANICAL SCULPTURE FROM *THE GUARDIAN OF THE TEMPLE*

written by Caleb Kraft

THE PEOPLE OF TOULOUSE, FRANCE WENT OUT INTO THE NOVEMBER MORNING TO FIND something very unusual in the street: a 46-foot-tall slumbering Minotaur, crouched and silent, the morning dew glistening off its hand-carved skin and steel bones.

The monster is the star of La Machine's modern interpretation of the classical tale of Ariadne (played by their giant spider, *La Princesse*). In this version, called *The Guardian of the Temple*, she is not an agent in the Minotaur's demise, but rather a protector, helping guide him to where he may find peace. Acted out over the course of several days, complete with live orchestral score, the company choreographed the two creatures' passage through the city. Crowds gathered to witness these two animated sculptures tell their tale.

La Machine, led by François Delarozière, calls itself a "street theatre company," but that description obscures the scale on which it operates. Its fantastic, moving creations — spiders, elephants, ants, and herons — tend to be many stories tall. While these are technically puppets, the team melds the most impressive elements of craftsmanship and engineering to build creations unlike anything seen before.

In addition to debuting their latest mechanical masterpiece, the November event celebrated a new Toulouse facility in which to house the colossal creations of La Machine. Typically the sculptures are crated and hidden away between events. Now, they will have a new home in the *Halle De La Machine*, where they can be on display and interactive year-round.

"I chose the Minotaur as a machine for Toulouse eight years ago," says Delarozière. "Toulouse is an ancient city and its city center, made of narrow, monochrome streets, looks like a labyrinth. Toulouse is also at the door of Spain and the bull is present in the history of the city."

On a 2016 visit to La Machine's workshop in Nantes, France, I got a sneak peek at the then-under-construction Minotaur.

CALEB KRAFT is a senior editor for *Make:* magazine. He has traveled the world documenting makers and their creations for the global community to enjoy.

Christian Gazzera

Mythical Magic

"I CHOSE THE MINOTAUR AS A MACHINE FOR TOULOUSE EIGHT YEARS AGO. TOULOUSE IS AN ANCIENT CITY AND ITS CITY CENTER, MADE OF NARROW, MONOCHROME STREETS, LOOKS LIKE A LABYRINTH." – *François Delarozière*

Most of the sculpted body parts were off the steel robotic structure that day, being finalized, finished, and detailed. The mechanical skeleton loomed over us. The sizable structure consists of roughly 50 tons of steel, hydraulic systems, and wood. You can truly feel that mass when it moves around you. Surprisingly, however, the movements are very graceful.

"The most complex part to develop was the exoskeleton," says Delarozière. "That allows the manipulation of the arms." It, along with the rest of the creation, is no simple matter. Like many of their works,

the Minotaur is steered not by an individual but by a team of artists, each piloting a separate appendage in unison with the others. They create an impressively cohesive movement, making the beast seem to walk and interact with the crowd.

What really surprised me about all the wooden parts is that they are hand sculpted to get to the final shape. Skilled artisans using power grinders and hand tools do all the finishing touches. The sheer quantity of sculpted parts was stunning, especially when you consider that there are multiples of these (in case something breaks), which

is true for all of La Machine's creations.

As a child, I read about the Seven Wonders of the World and thought to myself just how impressive it would have been to witness them while on an epic adventure. As I walked into La Machine's workshop, I was struck with the thought that this is what it must have been like.

As for their next build, it promises to be equally wonderful. "We are currently building a dragon for the city of Calais, the *Calais Dragon*," says Delarozière. "The beast will measure 25 meters long and weigh 70 tons." ✦

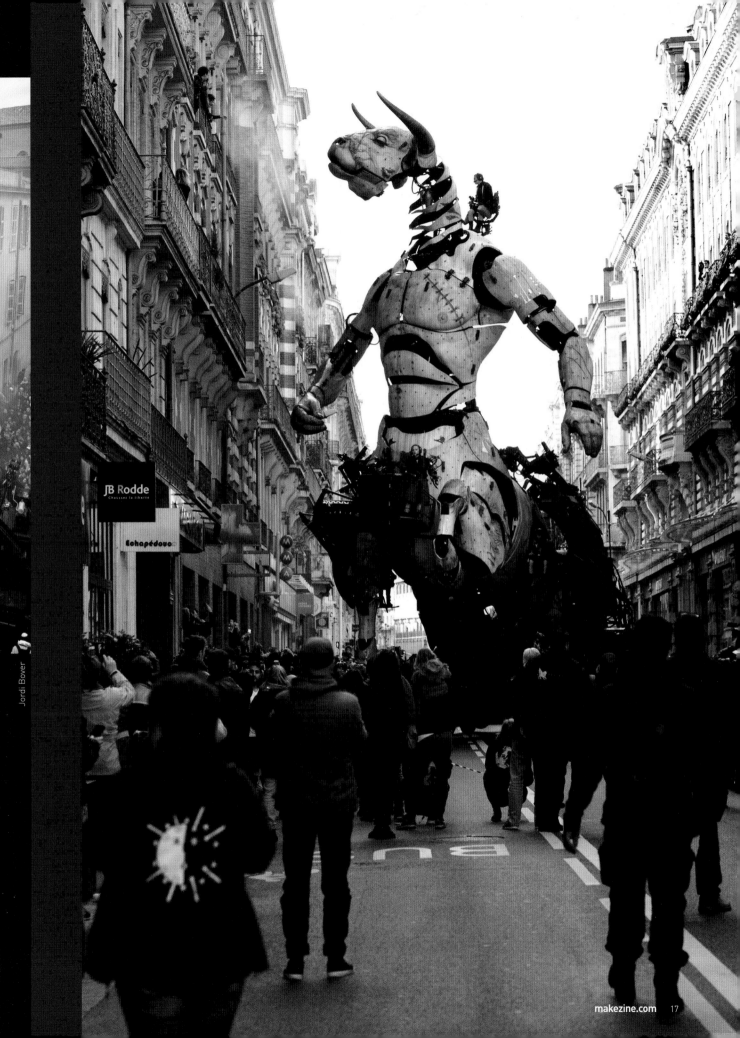

Jordi Bover

JB Rodde
Chaussez la liberté

Echapédoue

Mythical Magic

The Minotaur's skeleton loomed over us the day we visited as the majority of the body parts were off of the structure.

"THE MOST COMPLEX PART TO DEVELOP WAS THE EXOSKELETON. THAT ALLOWS THE MANIPULATION OF THE ARMS." – *François Delarozière*

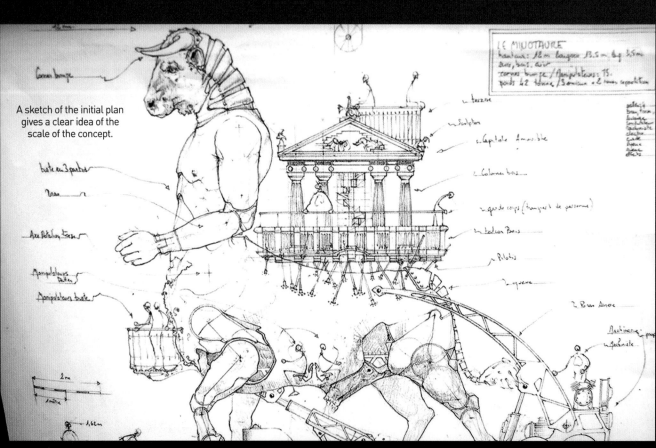

A sketch of the initial plan gives a clear idea of the scale of the concept.

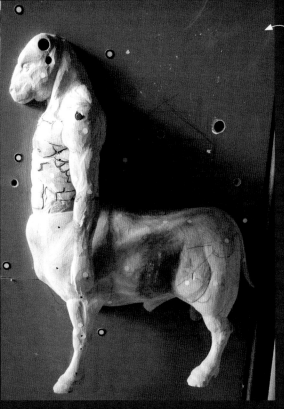

An early mock-up, where you can see the hand-drawn lines figuring out just how they should segment the body for movement.

François Delarozière, the man behind these constructions, has a wonderful, creative mind. I don't often get star-struck, but I bought one of their concept art books and had him sign it.

[+] See video, more photos, and behind-the-scenes looks at makezine.com/go/la-machine-minotaur

To control the left arm, a pilot dons this exoskeleton that allows the Minotaur to mimic the motion of the wearer.

Skilled artisans using power grinders and even hand tools to do final shaping.

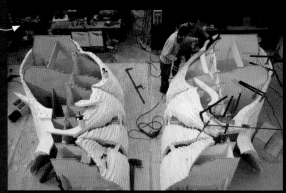

Caleb Kraft, Emmanuel Bourgeau

Making Time

THERE'S ONE RESOURCE WE JUST CAN'T GET ENOUGH OF, NO MATTER HOW HARD WE TRY

WRITTEN BY MIKE SENESE

TIME. IT'S OUR MOST PRECIOUS AND FINITE RESOURCE, ONE THAT WE'LL NEVER BE ABLE TO MANUFACTURE. We just can't get enough of it! What we can do, however, is make the most of what we've got.

It's a quandary that humans have obsessed over since the beginning, and it's no mistake that major developments in society connect to us mastering our use of time — both by being able to track it precisely (read about how the invention of accurate and portable watches and clocks greatly accelerated our understanding of the cosmos and the sciences on page 40), and by creating efficient systems that let us work — individually and as teams — in the best ways possible to accomplish tasks that would otherwise take forever (read our time-saving shop tips on page 36 and project management guidance on page 42).

This is an obsession for the tinkerers and engineers of the world. Some honor it by dedicating themselves to building their own clocks, from faithful reproductions of tiny, intricate, historic mechanisms all the way to futuristic reinventions of how time is tracked and displayed. Others pour themselves into those tools that let them manage time as best they can. Scientists study time's fascinating properties, fluctuating with exposure to gravity and velocity. And philosophers continue to ponder the meaning of it all.

And you, what will you make time for?

Time Will Tell

BUILD A MINIATURE NEOMATRIX WORD CLOCK
THAT USES SMART LEDS AND A GRID OF LETTERS
TO COLORFULLY SPELL OUT THE TIME

WRITTEN BY ANDY DORO

TIME REQUIRED:
2–3 Hours

DIFFICULTY:
Intermediate

COST:
$55–$65

MATERIALS
» **Pro Trinket 5V microcontroller** Adafruit #2000, adafruit.com
» **DS1307 real-time clock (RTC) breakout board** Adafruit #3296
» **NeoPixel NeoMatrix 8×8 RGB LED display** Adafruit #1487
» **Laser-cut enclosure** Find the files at github.com/andydoro/WordClock-NeoMatrix8x8 and cut them in ⅛" acrylic. I used clear for the box and black for the faceplate and pixel guard.
» **Machine screws, black nylon, #4-40 (14)**
» **Hex nuts, black nylon, #4-40 (14)**
» **Machine screws, #2-56 (2)** such as McMaster-Carr #91249A058, mcmaster.com
» **Hex nuts, #2-56 (4)**
» **Hookup wire, about 22–26 AWG** Silicone-covered wires are easiest to use but just about any wires will do.
» **Micro USB cable** such as Adafruit #2008
» **USB port power supply, 5V 1A (optional)** if you don't want to just power the clock from your computer, such as Adafruit #501
» **Plain paper (optional)** for diffuser

TOOLS
» **Computer with Arduino IDE** free from arduino.cc/downloads
» **Soldering iron and solder**
» **Wire strippers**
» **Diagonal cutters**
» **Slotted screwdriver, small (2.4mm)**
» **Tweezers**
» **Laser cutter (optional)** Cut the files yourself, or send them out to service for cutting.

ANDY DORO is an artist and designer from New York City. He works at Adafruit Industries as Director of Manufacturing and teaches physical computing at Hunter College.

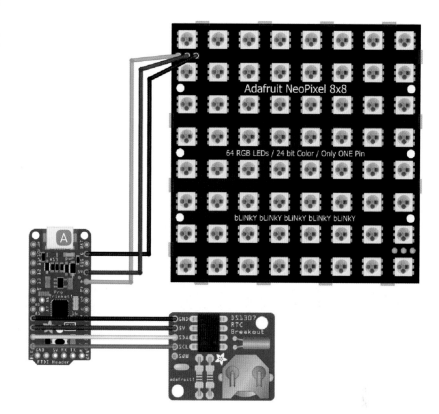

ARE YOU FASCINATED WITH THE PASSAGE OF TIME? Do you want a stylish, modern, and functional timepiece to add to your clock collection? The *word clock* is a one-of-a-kind time telling device, using a grid of letters to spell out the time. While you could spend thousands of dollars on other versions of this idea, this project is a quick and inexpensive way to build one for yourself.

This mini word clock uses the Adafruit NeoPixel NeoMatrix 8×8 display to create a colorful — and original — layout of letters that spell out the different time phrases. You can power it over USB so it makes for a great desk timekeeper. This clock also uses a DS1307 real-time clock breakout kit so it'll keep time even while unplugged! The DS1307 has an accuracy of +/– 2 seconds per day, and the word clock tells the time with a precision of 5 minutes: "five past," "ten past," "a quarter past," "twenty past," and so on. The microcontroller board we're using is the Adafruit Pro Trinket 5V but you can swap it with any Arduino-compatible microcontroller that can use I²C and NeoPixels.

[+] LEARN MORE ABOUT HOW TO USE REAL-TIME-CLOCKS (RTCs) IN OUR SKILL BUILDER ON PAGE 32.

CIRCUIT ASSEMBLY
The circuit is shown in Figure Ⓐ. We kept it compact by soldering the DS1307 RTC breakout directly onto the Pro Trinket 5V, using A2 and A3 as power pins (the DS1307 is powered by turning A2 low and A3 high in the Arduino code). This makes the reset button on the Pro Trinket 5V difficult to access, but you can still enter bootloader mode by using tweezers to push the button or just by unplugging and plugging the Pro Trinket 5V into the USB port; when it powers up, it will be in bootloader mode.

You don't have to solder them directly; you can also just free-wire the RTC breakout.

SOLDER THE RTC BREAKOUT BOARD
Start by assembling the DS1307 Real Time Clock breakout board by following the guide at learn.adafruit.com/ds1307-real-time-clock-breakout-board-kit. You only need to solder in the male headers for GND, 5V, SDA, and SCL. You can leave off SQW since it isn't used and the header won't fit nicely on top of the Pro Trinket. If you do solder it in, you can clip the bottom lead off.

Once the DS1307 breakout is assembled with headers, you can solder it on top of the Pro Trinket so that the DS1307 GND pin lines up with Trinket pin A2, 5V with A3, SDA

B

C

D

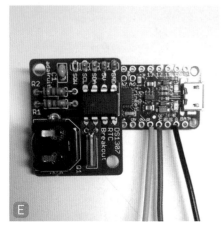

E

F

G

with A4, and SCL with A5 (Figure B). Make sure the boards are lined up correctly!

Solder the DS1307 RTC board onto the Pro Trinket 5V and then clip the excess leads (Figure C).

CONNECT THE NEOMATRIX 8×8 DISPLAY

Connect the NeoMatrix GND pin to the Pro Trinket GND, 5V to 5V, and DIN to Pin 8. Cut the wires to 5"–8" or 13cm–20cm long. Solder the wires into the back of the NeoMatrix so that they won't be visible from the front (Figure D).

Figure E shows the connections into the Pro Trinket 5V, and Figure F shows what the finished circuit should look like.

UPLOADING CODE

Make sure to use Arduino IDE 1.6.4 or higher and follow the tutorial at learn. adafruit.com/add-boards-arduino-v164 to install the Adafruit boards.

Download the code from github.com/ andydoro/WordClock-NeoMatrix8x8 by clicking Download ZIP. Uncompress the file and copy the folder to your Arduino sketchbook folder.

You will also need to install some Adafruit Arduino libraries: RTClib, DST_ RTC, Adafruit_GFX, Adafruit_NeoPixel, and Adafruit_NeoMatrix. (Follow the tutorial at learn.adafruit.com/adafruit-all-about-arduino-libraries-install-use if you're unfamiliar with how to do this.)

Once that's done, open the Arduino sketch *WordClock_NeoMatrix8x8.ino* in the Arduino IDE and select Tools→Board→Pro Trinket 5V/16MHz (USB) (Figure G). See if the code compiles. If the libraries aren't installed correctly you'll see errors.

Put the Pro Trinket into bootloader mode either by unplugging and replugging it into the computer with your micro USB cable or by hitting the reset button. Again, the reset button can be difficult to access if you've soldered the RTC on top, so I find plugging the board into USB to work best.

When the red LED on the Pro Trinket is pulsing, the board is in bootloader mode. Upload the code. If everything was done correctly, it should start telling you the time!

UNDERSTANDING THE CODE
STARTUP SEQUENCE

When the clock starts up, all of the individual words will light up sequentially. This startup sequence is defined by the flashWords() function in the setup() function.

Easter egg: I hid my "signature" in the letter grid. It lights up briefly during the startup sequence. :)

SETTING THE TIME
The first time the code is run, the time on the RTC module will be set to the time that the code was compiled on your computer.

```
if (! RTC.isrunning()) {
    Serial.println("RTC is NOT
running!");
    // following line sets the
RTC to the date & time this
sketch was compiled
    RTC.adjust(DateTime(__
DATE__, __TIME__));
    // DST? If we're in it,
let's subtract an hour from
the RTC time to keep our DST
calculation correct. This gives
us
    // Standard Time which our
DST check will add an hour back
to if we're in DST.
    DateTime standardTime = RTC.
now();
    if (dst_rtc.
checkDST(standardTime) == true)
{ // check whether we're in DST
right now. If we are, subtract
an hour.
        standardTime =
standardTime.unixtime() - 3600;
    }
    RTC.adjust(standardTime);
}
```

If you need to reset the time, this can be done by commenting out the `if` statement here but leaving in the line:

```
RTC.adjust(DateTime(__DATE__,
__TIME__));
```

Since this clock only spells out the time in 5-minute intervals, some find it helpful to add 2.5 minutes (or 150 seconds) to the actual time to give a closer account of the time:

```
// add 2.5 minutes to get better
estimates
    theTime = RTC.now();
    theTime = theTime.unixtime()
+ 150;
    RTC.adjust(theTime);
```

The last line adjusts the real-time clock. Very important!

DAYLIGHT SAVING TIME
Do you live in a territory that observes daylight saving time (DST)? If you do, you usually have to reprogram your clocks twice a year! This clock includes some code so that the adjustments are made automatically. The code follows the current rules for DST in the USA and Canada. If you live somewhere that follows different DST rules you may be able to modify the code to suit your rules — just look in the DST_RTC library functions. Wikipedia has a great reference at en.wikipedia.org/wiki/Daylight_saving_time_by_country.

If you live in a territory that doesn't observe daylight saving time, just alter the following line by changing the 1 to 0.

```
#define OBSERVE_DST 1
```

The daylight saving time code works by keeping the real-time clock on "standard time" and checking to see if the current date falls within daylight saving time. If the date falls within daylight saving time, an hour is added to the displayed time to convert from standard time to daylight saving time.

BRIGHTNESS ADJUSTMENT
The clock is programmed to change brightness automatically based on the time of day, operating at a lower brightness at nighttime. You can change these settings in the adjustBrightness tab.

DAYBRIGHTNESS is the brightness level during the day, and NIGHTBRIGHTNESS is the brightness level during the night. Any number between 0 and 255 is acceptable, but at 0 you won't see anything! 255 is maximum brightness.

MORNINGCUTOFF and NIGHTCUTOFF set which hours the clock will operate at DAYBRIGHTNESS and NIGHTBRIGHTNESS. These hours are for a 24-hour clock, so

22 would be 10pm. Between midnight and 1am, the hour is 0.

```
// brightness based on time of
day- could try warmer colors at
night?
#define DAYBRIGHTNESS 80
#define NIGHTBRIGHTNESS 40

// cutoff times for day / night
brightness. feel free to modify.
#define MORNINGCUTOFF 7  // when
does daybrightness begin?  7am
#define NIGHTCUTOFF   22 // when
does nightbrightness begin? 10pm
```

If you don't want the clock to change brightness throughout the day, you can also just comment out the adjustBrightness() function in the main loop().

COLOR SHIFTING SPEED
The speed at which the colors shift is controlled with a delay in milliseconds defined as SHIFTDELAY. To speed up the color shifting, decrease the number. To slow down the color shifting, increase the number.

```
#define SHIFTDELAY 100   //
controls color shifting speed
```

MONOCHROME MODE
Maybe the constant color shifting makes you queasy. You can simply run the clock to light the display in a single color.

First, we define that color — in this case, white:

```
// if you want to just run the
clock monochrome
#define WHITE 200, 255, 255
```

255, 255, 255 would be pure white, but with RGB LEDs this can look a little purple, so you can try adjusting the numbers.

Now in the colorFunctions tab, edit the two lines that set the PixelColor. Comment out the longer line and uncomment the shorter line.

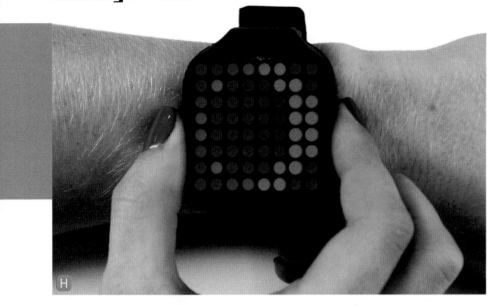

```
//matrix.setPixelColor(i,
Wheel(((i * 256 / matrix.
numPixels()) + j) & 255));
matrix.setPixelColor(i, WHITE);
```

MOON MODE

Moon Mode is something we borrowed from Phil Burgess' TimeSquare wristwatch (Figure H). If you feel like it, uncomment this line and your 8×8 Word Clock will show the current phase of the moon.

```
//mode_moon(); // uncomment to
show moon mode instead!
```

ENCLOSURE ASSEMBLY

Now that your circuit is complete, it's time to assemble the laser-cut enclosure. You'll need to find a laser-cutting shop, hackerspace, or other friend with a laser cutter to cut out the pieces. You can find the files to cut at github.com/andydoro/WordClock-NeoMatrix8x8. Use ⅛" clear and black acrylic — or get creative and do something else!

Start by attaching the NeoMatrix to the acrylic plate that will hold it inside the enclosure (Figure I). Four screws hold it in place (Figure J).

Now take the back panel and insert the #2-56 machine screws, which will hold the Pro Trinket in place (Figure K).

Attach the Pro Trinket to the back plate, using a pair of nuts as standoffs, and making sure the screws are tightened down firmly (Figure L).

Connect the NeoMatrix plate and the back panel to the side panel that has the hole for the micro USB.

Now you can add the other side panel and the top (Figure M) and bottom panels (Figure N), attaching each with the black nylon screws as you go.

Once all the clear acrylic pieces are put together, you're ready to add the NeoPixel guard and diffuser. Put the guard in place on top of the NeoMatrix (Figures O and P). This will help contain the light from each pixel, making each letter on your clock crisper and easier to read.

If you'd like to add a diffuser, now's the time. Diffusers are used to spread out the light from the NeoPixels and make the text

Top panel

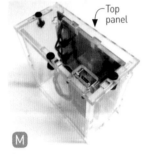

Bottom panel

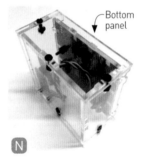

NeoMatrix

on the faceplate easier to read. You can make a diffuser from a plain sheet of paper, or any other material that will even out the bright light from the NeoPixels. Just trace the outline of the NeoMatrix and cut it out.

Place the diffuser on top of the pixel guard (Figure Q).

Now you're ready to attach the faceplate. First pull the protective paper cover off (Figure R) and use tweezers to poke out any remaining bits of the cut-out letters (Figure S).

Finally, put the faceplate onto the enclosure (Figure T), sealing the diffuser inside. Use the final four screws to affix the faceplate to the enclosure (Figure U).

TIME TO CELEBRATE

Your word clock is assembled! Revel in your accomplishment (Figure V).

And if you'd like to build an even tinier version, I also created a word clock version of the TimeSquare wristwatch (Figure W). Laser-cut or 3D-print the tiny faceplate and you can wear a word clock all day long. ◉

• • •

Special thanks to Dano Wall for designing the ingenious enclosure and faceplate, and for project inspiration! github.com/danowall

P

Q

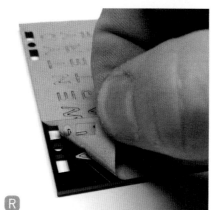
R

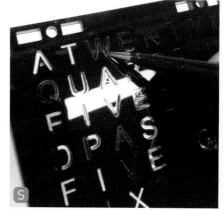
S

T

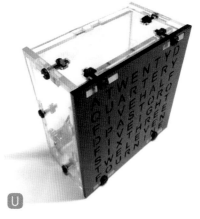
U

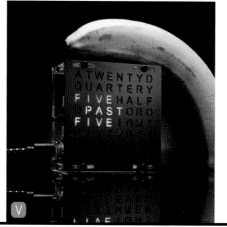
V

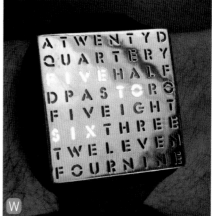
W

[+] FOLLOW THE GUIDES:

NEOMATRIX 8×8 WORD CLOCK
LEARN.ADAFRUIT.COM/
NEOMATRIX-8X8-WORD-CLOCK

TIMESQUARE WRISTWATCH WORD
CLOCK LEARN.ADAFRUIT.COM/
TIMESQUARE-WORDCLOCK

Time Tube

REIMAGINE THE
HOURGLASS —
WITH FALLING
MAGNETS SLOWED
BY MYSTERIOUS
EDDY CURRENTS

WRITTEN AND
PHOTOGRAPHED BY
CHARLES PLATT

CHARLES PLATT
is the author of
Make: Electronics, an
introductory guide for
all ages, its sequel
Make: More Electronics,
and the 3-volume
*Encyclopedia of
Electronic Components*.
His new book, *Make:
Tools*, is available now.
makershed.com/platt

TIME REQUIRED:
6–8 Hours

DIFFICULTY:
Intermediate

COST:
$20–$30

MATERIALS

» **Neodymium ball magnet, ⅝" diameter minimum** K&J Magnetics has a very wide range.

> **CAUTION:** Neodymium magnets can be dangerous. Large ones can move violently with enough force to break bones. Don't use them around magnetic objects of any kind.

» **Aluminum tube of internal diameter slightly larger than the ball magnet; length at least 9"** McMaster-Carr is a good source. See sidebar "Dropping the Ball" on page 31 for a discussion.

» **PVC plumbing elbows (2)** matching internal diameter of aluminum tube. Optionally, you can bend your own elbows from PVC pipe (see text).

» **Gearhead motor, 12VDC, about 1" diameter, 100rpm or slower**

» **AC adapter, 12VDC output, minimum 250mA** to power the motor

» **Reed switches, normally open, at least 100mA** Search eBay for all of these items or try Jameco Electronics or Electronics Goldmine.

» **Signal relays, DPDT 12VDC** such as NEC part #EC2-12NJ; from any electronics vendor

» **Diode, 1N4001** to suppress voltage spikes from the motor

» **Wood dowels:** ½" diameter, 6" length (1); 2" diameter, 1" length (1); and ¼" diameter, 12" length (1)

» **Pine boards:** 1×4, 24" length (1) and 2×6, 3" length (1)

» **Plywood, ¼"×12"×24"** If you increase the length of the aluminum tube, increase the dimensions of the wooden parts in ratio.

» **Rubber band or belt, about 4" long and at least ¼" wide**

» **Hookup wire**

» **Perforated board**

» **Epoxy glue**

» **Wood screws**

TOOLS

» **Handsaw** for wood, e.g. a tenon saw

» **Hacksaw** for aluminum

» **Drill with small hole saws** or Forstner bits, preferred

» **Screwdriver**

» **Soldering iron and solder**

OPTIONAL, TO CUT SLOT IN TUBE

» **Handheld circular saw with abrasive disc, clamps, scrap wood, and eye protection**

OPTIONAL, TO BEND PVC ELBOWS

» **Heat gun, extension spring, insulated gloves, fine sandpaper, and water spray** McMaster carries suitable extension springs.

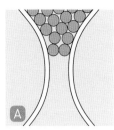
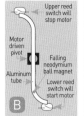
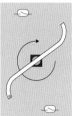

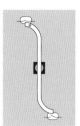

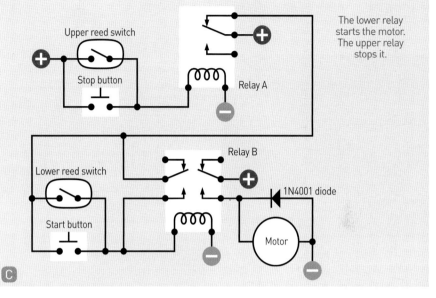

The lower relay starts the motor. The upper relay stops it.

AN HOURGLASS IS A BEAUTIFUL OBJECT, BUT IT HAS PROBLEMS.

If it runs a little too fast or slow, there's no way to adjust it. Sometimes the sand inside clogs up when two particles try to enter the neck side by side (Figure A), and you have to shake it to start the sand flowing again. Even when it's working right, a typical palm-sized hourglass has to be turned upside-down repeatedly to measure intervals beyond a couple of minutes.

I decided that a redesign was overdue.

CONCEPT

My first step was to change the name. I wasn't going to be measuring hours, and I wouldn't try to work with glass, so I called my gadget a Time Tube.

To prevent particles from clogging, I adopted the radical concept of using just one particle. Of course a single falling particle can only measure very brief intervals of time, but when you drop a magnet through a copper or aluminum tube, it falls surprisingly slowly as a result of creating eddy currents. (See "Dropping the Ball" on page 31.) Actually a ⅝" spherical neodymium magnet will descend at about 9 inches per second, after which the tube must still be turned

upside-down, hourglass-style — but I would use a motor to take care of that.

Figure B shows a diagram of the concept. As the ball reaches the bottom of the tube, its magnetic field closes the contacts of a reed switch. (You can learn more about reed switches in my book *Make: Electronics.*) The reed switch triggers a relay, which starts a motor. The motor rotates the tube 180°, and the magnetic ball trips another reed switch, which shuts off the motor. The ball falls through the tube again, and the cycle repeats.

Figure C shows the schematic. The lower reed switch, or the pushbutton in parallel with it, will energize Relay B, pulling its movable contacts downward. (Some relays move the contacts in the opposite direction — check your datasheet to make sure.) In my schematic, the right-hand contacts start the motor while the left-hand contacts cause the relay to continue energizing itself even when the reed switch opens.

The motor keeps running until the upper reed switch triggers Relay B, opening its contacts, which are normally closed. This shuts off Relay A, so the motor stops while the ball starts to fall.

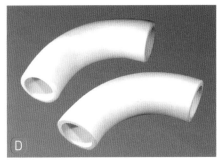

PVC pipe bent with a heat gun, cut to size, and smoothed with fine sandpaper.

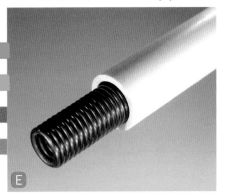

Before bending the pipe, insert a spring to prevent kinking.

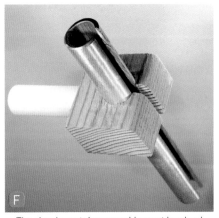

The aluminum tube assembly must be glued together, as any steel fasteners may stop the neodymium ball magnet from falling.

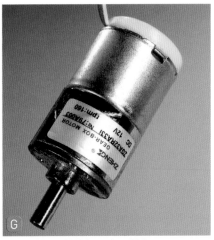

Use a 12VDC gearhead DC motor.

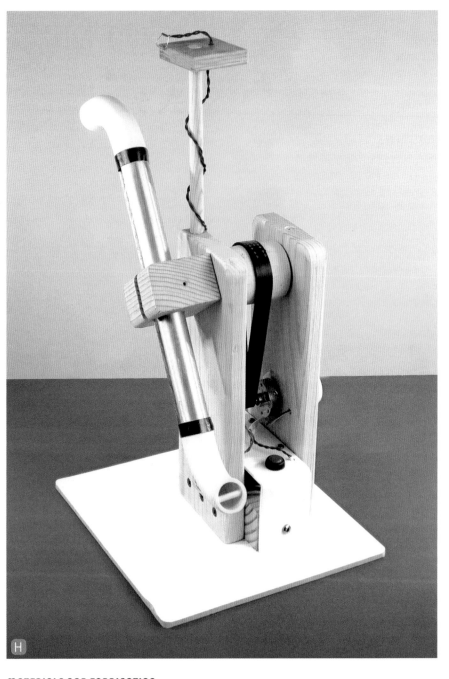

MATERIALS AND FABRICATION

I decided that my first Time Tube would just be a proof-of-concept design, to keep things simple. I wasn't sure of the optimal magnet size, so I chose one that's ⅝" in diameter, to fit inside an aluminum tube of 1" external diameter, with walls ⅛" thick, allowing a ¾" internal diameter. Tubing of these dimensions is very easy to find.

I used a piece of tube only 9" long, so that I wouldn't need an elaborate structure and a large motor. Also, a short tube would be easier to photograph.

I cut a slot along the length of the tube to reveal the ball as it falls. If you do this,

I suggest using an abrasive wheel on a handheld circular saw. Clamp two strips of wood horizontally on either side of the tube to support the saw while it slides along. Apply the clamps very firmly, wear eye protection, and do not use a wood-cutting blade! (Visit this project online at makezine. com/go/time-tube to see my cutting jig.)

Now you need to cap each end of the tube with a short piece of PVC pipe, curved so that when the tube rotates, the ball won't fall again until the tube is vertical. You can use off-the-shelf PVC elbows from the plumbing department of a hardware store, but I chose to bend my own (Figure D). To do this yourself, first insert a long extension spring in the pipe to stop it from kinking, as shown in Figure E. Apply a heat gun, turning the pipe frequently till it gets soft. Then bend it and squirt water on it to set it. Wear gloves that have some thermal insulation.

The falling magnetic ball will stop if it passes anything made of steel, so you can't mount the tube on a steel shaft. I inserted it in a hole in a block of wood and glued that to a shaft of low-friction nylon tube that I happened to have (Figure F). You can use wooden dowel instead, but you may have to lubricate it, in which case talcum powder works well.

I used a simple DC gearhead motor (Figure G). To increase its torque I connected it with a pulley made from 2" dowel, as shown in Figure H. The drive belt is from a vacuum cleaner, but a large rubber band may also do the job.

The electronics are housed under a rectangle of ⅛" ABS plastic or ¼" plywood. The reed switches are glued to small wood blocks, the upper being attached to a vertical length of ½" dowel. The pipe elbows can be capped, or blocked with a nonmagnetic rod or pin.

CALIBRATING YOUR TIME TUBE

If you enjoy the weirdness of the Time Tube but would like it to be just a little more practical, maybe you should add a digital readout. You can do this by triggering an Arduino with the output from one of the relays.

Suppose one full cycle of the Time Tube takes 3.5 seconds, or 7/120ths of a minute. To convert this to minutes, each time the Arduino receives a pulse from the relay it adds 7/120 to an internal floating-point variable, then copies the value to an integer variable, which sends its value to a digital display.

But wait. The Arduino has a perfectly good clock of its own, inside its hardware. So why do you even need a Time Tube?

Well, you don't. But watching an Arduino isn't nearly as much fun as watching a neodymium ball floating through a column of air. ●

[+] SHARE YOUR EXPERIMENTS
AND SHOW YOUR BUILD AT
MAKEZINE.COM/GO/TIME-TUBE.

Rob Nance

DROPPING THE BALL

A magnet falls slowly through an aluminum tube because the magnetic field has to do some work, creating **eddy currents** in the aluminum. I described this phenomenon in *Make:* Volume 59 (makezine.com/projects/explore-lenz-law) and used it to light LEDs. It's well known but difficult to measure, so you have to explore it using trial and error.

One thing we do know is that if you move twice as close to a point-sized magnet, it exerts four times as much force. This means that the slowing effect on the falling ball magnet will increase radically if you can find a tube that is only a fraction bigger than the ball.

When you start shopping for aluminum tube, you'll often find it described by its outside diameter, abbreviated OD and measured in fractions of an inch, while the wall thickness is measured in decimal values of an inch, and you have to figure out the internal diameter (ID) yourself.

I'll use my test tube as an example.
» OD: 1"
» Wall thickness: ⅛" = 0.125"
» Subtract double the wall thickness from OD to get ID: 1.0 − 0.125 − 0.125 = 0.75"
» Ball diameter: ⅝" = 0.625"
» Total gap around the ball: ID minus ball diameter, which is: 0.75 − 0.625 = 0.125"

Shopping online I found some aluminum tube with OD ¾" (0.75") and walls only 0.028" thick. So the ID was 0.75 − 0.028 − 0.028 = 0.694" and therefore the total gap around the ball was 0.694 − 0.625 = 0.069". When I dropped the ball through this tube, it took a full 5 seconds to fall 3 feet.

Could I do better than that? Yes, I found a piece of tube with ¾" OD but thicker walls, allowing less room inside. The walls were 0.04" thick, so the internal diameter was 0.75 − 0.04 − 0.04 = 0.67" and the total gap around the ball was 0.67 − 0.625 = 0.045". This time the ball took 8 seconds to fall 3 feet!

Objects falling freely under the force of gravity will accelerate, but a ball magnet resisting gravity seems to fall at a constant speed. Therefore, if you double the length of the tube, the ball should take twice as long to reach the bottom.

How much will the speed change when you cut a slot in the tube? I don't know. If it does change, will the width of the slot make a big difference? I don't know. Will bigger ball magnets perform better than small ones? I don't know that, either! The mass of the ball will increase with the cube of its diameter, but who knows, its magnetic field may increase by that much too.

You'll need to do some experimenting to find out. But bear in mind that a longer, fatter tube and a heavier magnet will require a much more powerful motor to turn the tube upside-down while the ball is sitting at one end.

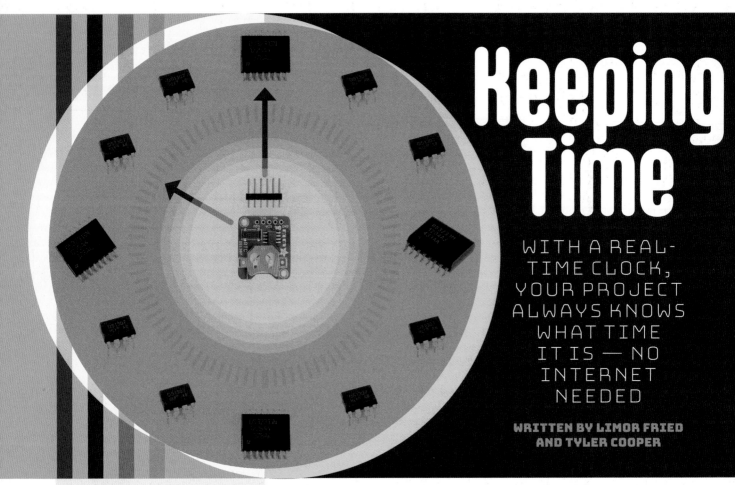

Keeping Time

WITH A REAL-TIME CLOCK, YOUR PROJECT ALWAYS KNOWS WHAT TIME IT IS — NO INTERNET NEEDED

WRITTEN BY LIMOR FRIED AND TYLER COOPER

LIMOR FRIED ("LADYADA") is the founder and lead engineer of Adafruit Industries.

TYLER COOPER is a creative engineer at Adafruit, developer of the Adafruit Learning System, and co-author of *Getting Started with Adafruit FLORA*.

A *REAL-TIME CLOCK (RTC)* CHIP IS **BASICALLY JUST LIKE A WATCH** — it runs on a battery and keeps time for you. Using an RTC in your project, you can keep track of long timelines, even if you reprogram your microcontroller or disconnect it from power.

Most microcontrollers, including the Arduino, have a built-in timekeeper function called `millis()`, as well as a timer built into their chips that can track longer periods like minutes or days. So why would you want a separate RTC chip? Well, `millis()` only keeps track of time since the Arduino was last powered on and the millisecond timer was set back to 0. The Arduino doesn't know it's "Tuesday" or "March 8th," only "It's been 14,000 milliseconds since I was turned on."

While this is OK for some projects, others such as data loggers, clocks, and alarms need to have consistent timekeeping that doesn't reset when the Arduino reboots. Thus, we include a separate RTC — a specialized chip that just keeps track of time. It counts leap years and knows how many days are in each month. Just note that it doesn't do Daylight Saving Time — you'll have to code that for your specific region.

CHOOSING AN RTC

Three RTCs are commonly used by makers. Each communicates via the two-wire I²C protocol and will merrily tick along for years on a coin cell.

» **PCF8523** — May lose or gain 2 seconds a day, but it's the least expensive and it works with 3.3V *or* 5V power and logic.

» **DS1307** — Also not high-precision, but it's low-cost and easy to solder. Requires 5V power but can work with 3.3V logic.

» **DS3231** — Temperature-compensated for extreme accuracy; works with 3.3V or 5V.

USING THE DS3231

Most RTCs use an external **32kHz timing crystal** to keep time, but those crystals can drift — the temperature affects the oscillation frequency very, very slightly but it does add up. The DS3231 is in a beefy package because the crystal is inside the chip, and right next to it is a **temperature sensor** that compensates for frequency changes by adding or removing clock ticks so that the timekeeping stays on schedule. Adafruit offers this RTC in a breadboard-friendly breakout board (Figure A).

DS3231 PINOUT

POWER PINS:

» **Vin** — Since the RTC can be powered by 2.3V–5.5V, you don't need a regulator or level shifter, just give this board the same power as your logic level.

» **GND** — Common ground, power and logic.

I2C LOGIC PINS:

» **SCL** — I²C clock pin, connect to your microcontroller's I²C clock line. This pin has a 10K pull-up resistor to Vin.

» **SDA** — I²C data, connect to microcontroller I²C data line. Has 10K pull-up.

OTHER PINS:

» **BAT** — Connects to positive pad of the battery. Use this if you want to power something else from the coin cell, or provide backup from a separate battery. VBat can be 2.3V–5.5V and the DS3231 will switch over when Vin power is lost.

» **32K** — The 32kHz oscillator output. Open drain, so you need to attach a pull-up to read it from a microcontroller pin.

» **SQW** — Square wave or interrupt output. Again, open drain, attach a pull-up.

» **RST** — Can reset an external device or indicate when main power is lost. Open drain, but has an internal 50K pull-up that keeps pin voltage high as long as Vin is present. When Vin drops and the chip switches to battery, this pin goes low.

ARDUINO USAGE

You can easily wire this breakout to any microcontroller; here we'll use an Arduino (Figure 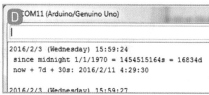). For other microcontrollers, just make sure it has I²C, then port the code.

» Connect Vin to the power supply, 3V–5V is fine. Again, use the same voltage as your microcontroller logic.

» Connect GND to common power/data ground.

» Connect the SCL pin to the I²C clock SCL pin on your Arduino.

» Connect the SDA pin to the I²C data SDA pin on your Arduino.

> **IMPORTANT:** The DS3231 has a default I²C address of **0x68** and cannot be changed.

YOUR FIRST RTC TEST

Now we'll demonstrate a test sketch that will read the time from the RTC once per second. To start, remove the battery from the holder while the Arduino is not powered, wait 3 seconds, and then replace the battery. This resets the RTC chip.

LOAD DEMO SKETCH

First download Adafruit's RTClib (a fork of JeeLab's excellent RTClib) from our GitHub repository, following the instructions at github.com/adafruit/RTClib.

Launch the Arduino IDE, open up File→Examples→RTClib→ds3231, and upload it to your Arduino.

Now check the Serial Monitor console at 9600 baud. After a few seconds, you'll see the report that the Arduino noticed this is the first time the DS3231 has been powered up, and will set the time based on the Arduino sketch (Figure 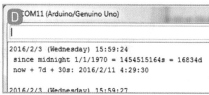).

Unplug your Arduino plus RTC for a few seconds (or minutes, or hours, or weeks) and power it up again. This time you won't get the "RTC lost power" message — instead it will immediately let you know the correct time (Figure 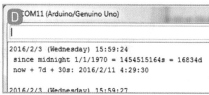)! You won't have to set the time again; the battery lasts 5 years.

READING THE TIME

Now we'll query the RTC for the time. Let's look at the sketch to see how it's done.

```
void loop () {
    DateTime now = rtc.now();

    Serial.print(now.year(), DEC);
    Serial.print('/');
    Serial.print(now.month(), DEC);
    Serial.print('/');
    Serial.print(now.day(), DEC);
    Serial.print(" (");
    Serial.print(daysOfTheWeek[now.
      dayOfTheWeek()]);
    Serial.print(") ");
    Serial.print(now.hour(), DEC);
    Serial.print(':');
    Serial.print(now.minute(), DEC);
    Serial.print(':');
    Serial.print(now.second(), DEC);
    Serial.println();
```

To get the time using RTClib, you call **now()**, a function that returns a **DateTime** object that describes the year, month, day, hour, minute, and second you called it.

Some RTC libraries instead have you call **RTC.year()** and **RTC.hour()**, but that method has a problem: If you happen to ask for the minute right at 3:14:59 and then again 1 second later (at 3:15:00) you'll see the time as **3:14:00** which is a minute off!

Instead, we take a "snapshot" of the time all at once, then pull it apart into **day()** or

We removed the power wire from Arduino 5V pin to the breadboard Vin rail before taking this pic, don't forget it!

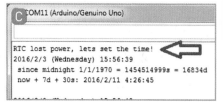

```
COM11 (Arduino/Genuino Uno)

RTC lost power, lets set the time!  ⟵
2016/2/3 (Wednesday) 15:56:39
  since midnight 1/1/1970 = 1454514999s = 16834d
  now + 7d + 30s: 2016/2/11 4:26:45
```

```
COM11 (Arduino/Genuino Uno)

2016/2/3 (Wednesday) 15:59:24
  since midnight 1/1/1970 = 1454515164s = 16834d
  now + 7d + 30s: 2016/2/11 4:29:30
2016/2/3 (Wednesday) 15:59:27
```

second() as seen above. It's a tiny bit more effort but it's worth it to avoid mistakes.

We can also get a "timestamp" out of the **DateTime** object by calling **unixtime()** which counts the seconds (not counting leap seconds) since midnight, January 1, 1970:

```
Serial.print(" since midnight
  1/1/1970 = ");
Serial.print(now.unixtime());
Serial.print("s = ");
Serial.print(now.unixtime() /
  86400L);
Serial.println("d");
```

Since there are 60 * 60 * 24 = 86,400 seconds in a day, we can easily count days too. This method can also make the math easier: To check if 5 minutes have passed, just see if **unixtime()** has increased by 300 and don't worry about hour changes. ●

[+] RASPBERRY PI? CIRCUIT-PYTHON? LEARN MORE AT MAKEZINE.COM/GO/SKILL-BUILDER-RTC.

Time Warp

THESE SUBLIMELY CREATIVE PROJECTS WILL LET YOU TELL TIME IN A TOTALLY DIFFERENT WAY

WRITTEN BY CALEB KRAFT

Clocks. They're so daily, so integral in our lives that they get constant homage in the form of unique and interesting design approaches. These following makers have taken the concept of keeping time and added a twist that makes them especially interesting.

CALEB KRAFT is a senior editor for *Make:* magazine. Working from home for years, he only views clocks as vaguely helpful artistic machines thanks to his flexible schedule.

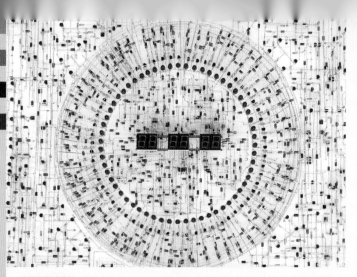

THE CLOCK

techno-logic-art.com/clock.htm

Rather than following the typical function-over-form approach to circuit design, Gislain Benoit went a completely different direction with this clock build, employing the "dead bug" approach to prototyping — soldering the parts to one another without a PCB — in an artistic style that makes this circuit into an awe inspiring sculpture. The work as a whole is both stunning and functional.

3D PRINTED TOURBILLON

makezine.com/projects/3d-printed-tourbillon-clock

If you were to go shopping for a wristwatch at the fanciest retailer you could think of, you would likely find at least one watch with a visible tourbillon, the internal mechanism that makes things tick. This version from Christoph Laimer is fully 3D printable and consists of over 50 parts. It comprises all the gears and pieces to actually function, including printed springs. Even if not used to tell the time of day, this is a must-have timepiece for any enthusiast.

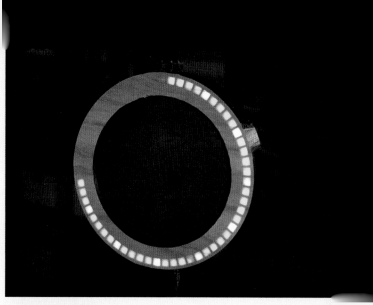

PARISIAN BINARY CLOCK

makezine.com/2017/06/03/model-parisian-building-binary-clock

It has always been a point of geek pride to be able to decipher the time in binary. Typically, this manifests itself in bare circuits with lopsided hand-soldered LED displays, or at best, stark designs with simple dots visible on the face. However, this beautiful timepiece eschews those common designs in favor of a Parisian facade.

This laser-cut miniature building, made by Lucas, Victor, and Philippe Berbesson, along with Claire Protin, will tell you the time by lighting the interiors of different rooms. A glance will tell you what time of day it is, as long as you can decipher the code. Even if you can't you'll probably enjoy the look of this artistic piece.

THE CLOCK TARGET

makezine.com/2017/02/20/nerf-dart-clock

Admit it — you've always wanted to pelt your alarm clock to make it stop buzzing in the morning. When creating a new timepiece for his home, Christopher Guichet went for a stylish design, putting a ring of hidden lights underneath the clock's thin wooden veneer, using various patterns and colors to display the time through the wood. And then he took it further by adding an impact-detection function using an Arduino and a piezo sensor. Now, when he wants to snooze a little longer, he can just take aim and blast it with a Nerf dart. Even if he misses, he's probably happy to have such a swank clock on his wall.

DOTTIE THE FLIP DOT CLOCK

dhenshaw.net/art/Dottie

Dottie is a project made by David Henshaw that repurposes an old flip dot display salvaged from a bus to tell the time, temperature, and date. The project involved reverse engineering the sign in order to design new circuits to handle not only keeping time, but actually driving the individual dots actions. He even went a step further to add chimes, giving this clock the ability to announce alarms.

DIGITAL SUNDIAL

makezine.com/2015/11/11/sundial-casts-digital-display-no-electronics-needed

In a visually stunning feat of engineering and math, this clock gives you a digital display using shadows from the sun itself! The Gnomon, as the creator Mojoptix calls it, has hundreds of tiny holes at perfect angles to allow sunlight to shine through, creating a pixelated pattern that turns into a digital-esque time display once every 20 minutes.

John Edgar Park recommends acetate overlays and dry erase markers over drawings to plan wiring, additional parts, etc.

Up Your Efficiency

TIME-SAVING TIPS AND TRICKS FROM EXPERT MAKERS TO HELP YOU GET THE MOST OUT OF YOUR WORK AND WORKSHOP

WRITTEN BY GARETH BRANWYN

GARETH BRANWYN is the former editorial director of *Make:* magazine, and a pioneer of both online culture and the maker movement. He is the author and editor of over a dozen books, and is currently a regular contributor to *Make:*, Boing Boing, and other online and offline publications.

STAYING ABREAST OF SHOP TIPS AND PROCEDURES CAN GO A LONG WAY toward making your project work more efficient and enjoyable. Here is a collection of time-savers from my book, *Tips and Tales from the Workshop*, contributed by expert makers and myself.

• • •

START YOUR WEEK ORGANIZED
One of the few organizational rituals that I follow is Sunday cleaning and organizing. I like to start the week with at least some semblance of organization by cleaning up my shop, organizing the papers on my desk, and thinking about what I have ahead for the week. This hour or two each week at least prevents the chaos around me from becoming too unmanageable.

ORGANIZE FOR FIRST ORDER RETRIEVABILITY
This can help reduce time to find and get your tools and materials. Arrange your workspace so that the more commonly used the tool or material is, the closer it is to you. Conversely, more occasional tools are farther away. This way, the shop is designed so that you can easily find what you need as you need it. — *Adam Savage*

USE SEE-THROUGH BINS FOR SHOP ORGANIZATION
Use clear bins in your shop to organize materials, tools, and supplies so that you can see at a glance what's in them. Organize items by use rather than material type. So, for instance, everything sanding-related might go into a bin, instead of sandpaper in one, sanding blocks in another, etc. — *Bill Livolsi*

ORDER MORE THAN YOU NEED
Always order 10% more materials than you need for a project. And if you're ordering cheap parts and supplies, always order a few extra. — *Tim Slagle*

YOUR HAND RULES!
Know the measurements of certain parts of your body (actual length of your foot, finger, or the span of your hand) for doing rough measurements when you don't have a ruler handy. — *Jimmy Diresta*

KEEPING TRACK OF SMALL PARTS
Use double-sided tape to hold small parts in place while you disassemble or reassemble something. Affix the tape to a piece of paper and write where the part goes.

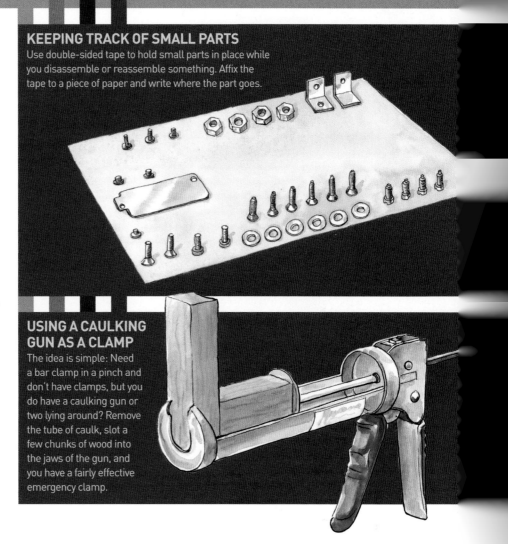

USING A CAULKING GUN AS A CLAMP
The idea is simple: Need a bar clamp in a pinch and don't have clamps, but you do have a caulking gun or two lying around? Remove the tube of caulk, slot a few chunks of wood into the jaws of the gun, and you have a fairly effective emergency clamp.

TRIPLE TIME
Estimate the time, expense, and number of supply runs required for a project intuitively. Then triple it, and plan accordingly. — *Sean Michael Ragan*

CREATE QUICK MEASURING STICKS
If you do a lot of woodworking and make many of the same cuts over and over, create some measuring sticks for the common lengths that you cut. Attach a large washer on one end that overhangs the edge to form a stop for holding the stick in place while you are scribing off the length. — *Dirt Farmer Jay*

QUICK-DRAWING LINES CLOSE TO THE EDGE
To draw a straight line close to the edge of a workpiece, hold your pencil in your hand and keep your fingers in exactly the same position while you follow the edge with the tip of your finger along the workpiece as you mark the line. — *JD*

QUICK-DRAWING LINES FARTHER FROM THE EDGE
If you need to draw a series of parallel lines farther from the edge of your workpiece, use a yardstick, ruler, or a piece of wood. Hold your pencil to the ruler at the desired distance from the edge and use your other hand to hold the ruler against the edge of the workpiece as you run the line. A tape measure works for this, too, but it gets more difficult the farther away you get from the edge. — *JD*

SHARPENED TWO-TIPPED PENCILS EVERYWHERE
Buy a box of eraserless pencils, sharpen both sides, and leave pencils at every work area of your shop. You'll always find a pencil, and it will likely have at least one sharp point on it. — *Dave Picciuto*

SCORE AND BREAK FOAMBOARD
If you have a lot of foamboard to cut down, rather than having to saw all the way

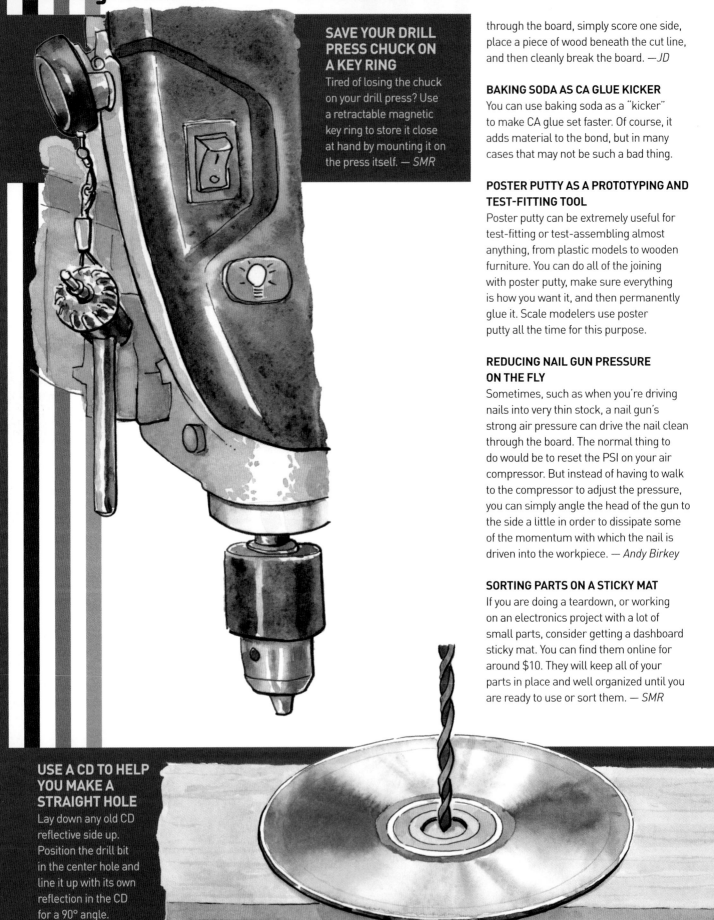

SAVE YOUR DRILL PRESS CHUCK ON A KEY RING

Tired of losing the chuck on your drill press? Use a retractable magnetic key ring to store it close at hand by mounting it on the press itself. — SMR

through the board, simply score one side, place a piece of wood beneath the cut line, and then cleanly break the board. —JD

BAKING SODA AS CA GLUE KICKER

You can use baking soda as a "kicker" to make CA glue set faster. Of course, it adds material to the bond, but in many cases that may not be such a bad thing.

POSTER PUTTY AS A PROTOTYPING AND TEST-FITTING TOOL

Poster putty can be extremely useful for test-fitting or test-assembling almost anything, from plastic models to wooden furniture. You can do all of the joining with poster putty, make sure everything is how you want it, and then permanently glue it. Scale modelers use poster putty all the time for this purpose.

REDUCING NAIL GUN PRESSURE ON THE FLY

Sometimes, such as when you're driving nails into very thin stock, a nail gun's strong air pressure can drive the nail clean through the board. The normal thing to do would be to reset the PSI on your air compressor. But instead of having to walk to the compressor to adjust the pressure, you can simply angle the head of the gun to the side a little in order to dissipate some of the momentum with which the nail is driven into the workpiece. — Andy Birkey

SORTING PARTS ON A STICKY MAT

If you are doing a teardown, or working on an electronics project with a lot of small parts, consider getting a dashboard sticky mat. You can find them online for around $10. They will keep all of your parts in place and well organized until you are ready to use or sort them. — SMR

USE A CD TO HELP YOU MAKE A STRAIGHT HOLE

Lay down any old CD reflective side up. Position the drill bit in the center hole and line it up with its own reflection in the CD for a 90° angle.

DEPOPULATE A PCB QUICKLY

Want to quickly desolder (aka, depopulate) a through-hole printed circuit board? Place it component-side down over a container and use a heat gun on the solder side. The components will quickly fall away into the container below. — *Scott Haun*

USING LEGOS FOR MOLD BOXES

If you've looked at any molding and casting how-tos online, or done any yourself, you likely already know this trick, but it's still worth mentioning. Lego bricks make for a perfect, reusable, and resizable mold box, and nearly every hobbyist (and pro) who does casting uses them.

USING A GLOVE TO PRESERVE A WORKING BRUSH

You can temporarily cover a paintbrush or stain brush by inverting your disposable glove over the brush between coats. You can also stretch the other glove over the can. This allows you to take a break without having to clean the brush or reseal the paint can.

MASKING WITH ALUMINUM FOIL

Use aluminum foil as a quick and clean masking material for doorknobs and fixtures.

USE A CERAMIC MUG AS A SHARPENING STONE

What can you do if you find yourself needing to sharpen a utility knife, pocket blade, or scissors and you don't have access to a sharpening stone? You can flip over a ceramic coffee mug and use the outer edge of the bottom (the non-glazed part) as an emergency sharpening stone. Try it. It works! — *Izzy Swan*

GO AS LOW RES AS YOU CAN

Before you decide on the resolution of your 3D print, think about the print's application. If you don't need a high-resolution print, try a lower-resolution setting. It will save you time, material, and wear and tear on your machine. — *Bob Clagett*

SKIPPING THE TABS ON CNC PRINTS

When using double-sided tape to secure your piece to the worktable, arrange the tape in a herringbone pattern. This way, you will "catch" the pieces that you're cutting out so that you don't have to include tabs in

SHORTEN A BOLT WITH A DRILL

From Family Handyman comes this tip: "If you need to shorten a bolt, let your drill do the hard work. Spin two nuts onto the bolt, tightening them against each other. Then, chuck the bolt into the drill and hold a hacksaw blade against the spinning bolt. The nuts help to steady the blade and clean off burrs when you unscrew them."

DIY CONTAINMENT BOX

You can easily make a simple containment box to prevent dust and small particles from escaping into the room when you're sanding small items. Basically all you need is a cardboard box, some old rubber gloves, a few layers of plastic wrap, and masking tape.

your design. After the machine is finished, carefully pry away the waste material, and your cut pieces will remain stuck to the tape and the worktable where they can be carefully pried up. — *Josh Price*

THREAD A NEEDLE THE EASY WAY

Here's a very clever method of threading a needle. Basically, you place the thread across the palm of your hand, place the eye of the needle perpendicular to the thread, press the needle into the thread, and wiggle it back and forth across the thread until a loop of thread gets worked through the eye. Pull on that loop and you are done. It's useful (if you don't have a needle threader

handy), and it feels kind of like a magic trick at the same time.

PHOTO-DOCUMENT PROJECTS IN REVERSE

While taking apart an item, I shoot a photo of each step. If I remove a screw, I take a photo. If I take spring slides off a pin, I take a photo. Once the object is fully disassembled, you simply reverse the order of the photos and you have a visual representation of how to put the thing back together. And it helps to make sure you write the steps accurately, because every step has been visually documented. — *James Floyd Kelly*

Richard Sheppard

Poetry in Motion

EXPLORING THE MAGIC OF MECHANICAL TIMEKEEPING

WRITTEN BY JORDAN FICKLIN

JORDAN FICKLIN has been working with watches and clocks since 2001 and is the Executive Director of the American Watchmakers-Clockmakers Institute.

DO YOU WANT TO GET STARTED MAKING CLOCKS OR WATCHES?

» Participate in a *Build a Watch* or *Build a Clock* class with the American Watchmakers-Clockmakers Institute (AWCI) awci.com/buildawatch

» Become a member of AWCI or the National Association of Watch and Clock Collectors (NAWCC) to gain access to a wealth of resources including horological libraries, magazines, videos, and a worldwide network of horological makers who meet regularly at local chapter meetings.

» Start with a simple kit from clockkit.com

IT WASN'T UNTIL THE MIDDLE OF THE 17ᵗʰ CENTURY THAT WE REALLY STARTED to see clocks as we know them. Around this time Galileo proposed using a pendulum to regulate a clock and Christiaan Huygens built a mechanical clock with a pendulum. Clocks quickly became accurate enough to add a minute hand and with additional improvements to compensate for changes in temperature and pressure, a second hand. For almost 300 years the pendulum clock was the most accurate method for telling time. Accurate clocks have harkened discoveries ranging from astronomical measurement to Einstein's theories.

Not long after the invention of the pendulum clock, Robert Hooke and Huygens developed portable clocks, called watches, which utilized a balance wheel with a spring for regulation. With improvements by great scientists and watchmakers, watches became essential tools for navigation, saving millions of lives. John Harrison's chronometer solved the problem of determining longitude at sea; American-made railroad pocket watches combined with standardized time zones kept trains from colliding.

INTRODUCING OSCILLATIONS

In the 20th century, advances in measuring time progressed much faster than they had in the four centuries prior. The discovery of the piezoelectric nature of quartz crystals (by Marie Curie's husband, Pierre, and his brother Jacques) cleared the path for electric timepieces, which relied on the regular oscillation of a quartz crystal. The first quartz clock was developed in 1927 with a level of accuracy that rivaled that of a pendulum regulator. The atomic era gave way to atomic clocks, which relied upon observing the regular oscillation of a cesium atom. These clocks were so precise that the second had to be redefined because scientists determined that the Earth's rotation was slowing. In 1967 the second was no longer 1/86,400 of a mean solar day. It's new definition: the duration of 9,192,631,770 periods of the radiation corresponding to the transition between the two hyperfine levels of the ground state of the cesium 133 atom.

A TUG OF WAR

Despite these incredible advancements in timekeeping, the mechanical clock and wristwatch are still respected as valuable machines. There is something magical about a tiny mechanical machine, especially one that is handmade.

These timepieces are poetic. They are art. At the heart of any mechanical clock or watch is the escapement. A clock is a tug of war between two opposing forces, each one giving and taking just the right amount to keep the clock running and on time. Ultimately the power comes from the

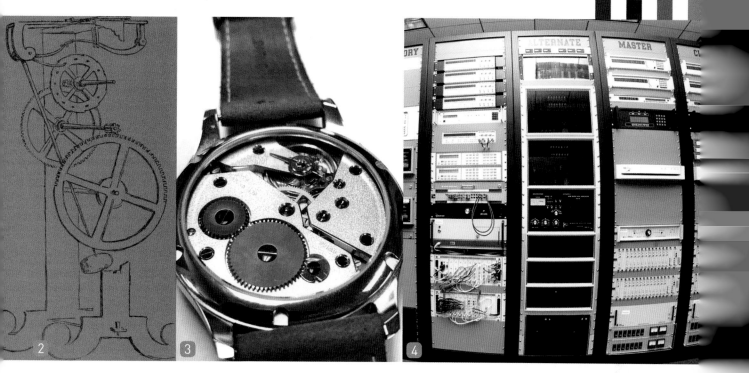

user. When you wind your clock you store your energy in the machine. It may be a coiled spring with the potential to unwind or a risen weight with the potential to drop. Either way the clock has captured your energy. Over the course of a day or a week the spring slowly unwinds, or the weight slowly falls, releasing your energy in a very controlled fashion.

The energy is transferred through a system of tiny, precise, brass gears to the escapement. The escapement, as its name suggests, holds back the energy and allows it to be released (or escape) in a controlled fashion. With each swing of the pendulum a tiny amount of energy is released. The gears advance and the pendulum is sent swinging away from its lowest point, fighting the force of gravity. Ultimately the force of gravity overcomes the pendulum and brings it back down to its lowest point, where the pendulum unlocks the escapement and allows a tiny amount more of energy to flow from the clock. The escapement gives the pendulum a little impulse, sending it on its way in the opposite direction as before. This delicate dance continues until all the energy stored in the clock has been lost to friction or released through the escapement into the pendulum.

REFINED PRECISION

Each of these mechanical components is subject to all kinds of influences, which might introduce errors. They expand and contract with temperature changes. Changes in barometric pressure affect the resistance on the pendulum offered by the atmosphere. Changes in the lubrication affect the amount of energy transferred through the system. For watches you have the added challenges of more extreme changes in temperature because the watch changes environments, not to mention the simple fact that it is constantly being moved around and jolted.

Despite all of this, these machines can be more than precise enough for most of our needs. A well-made regulator clock with compensating pendulum can keep time to within 1 second a month (better than your average quartz watch). A chronometer-grade wristwatch is accurate to within about a minute a month. The horologists of the world (watchmakers and clockmakers) stand in awe at these tiny machines and strive to make them better.

The art is both in the precision of the operation and in the careful decoration. Watches and clocks are art. Handmade timepieces are unique and craftsmen pour hundreds and sometimes thousands of hours into making each part both function perfectly and look beautiful. And with a world where the precise time can be found on dozens of devices in every room we still value these machines, perhaps because they don't really keep time, but they keep up with it.

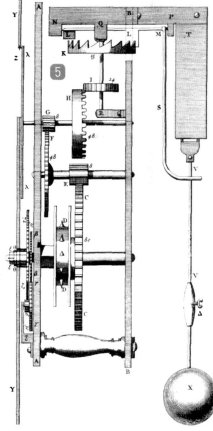

1 A 19th-century great-wheel skeleton clock.

2 Galileo's illustrated pendulum clock concept.

3 A mechanical watch I built.

4 U.S. Naval Observatory Alternate Master Clock.

5 Christiaan Huygens' pendulum clock.

Successful Solutions

FROM SMALL TASKS TO BIG, PROJECT MANAGEMENT TOOLS WILL HELP KEEP YOU ORGANIZED AND SAVE TIME

WRITTEN BY
JAKE GUTTORMSSON

WHETHER IT'S WOODWORKING IN YOUR BASEMENT OR BUILDING THE NEXT MARS PROBE, everyone needs project management. If you plan out your timelines, tasks, and milestones well, you can enjoy your work without stress, fatigue, or guilt. The basics of project management are simple: break the work into tasks, determine the time needed for each task, and identify the resources needed for each. Fortunately, software solutions provide ample organization for all this.

Two key notes: Individually, it takes **discipline** to find a system, stick to it, and finish each task in a thoughtful and timely manner. And collectively, it takes a lot of **communication**.

SMALL STUFF

For beginners and most individual projects, you can use something like a spreadsheet in Google docs. Beyond just a to-do list, you can group and sort tasks, update statuses, record notes, and collaborate with others. I also like Airtable for personal work — it takes a Google doc-like spreadsheet, and turns it on its head, allowing you to filter the data and view it as a grid, calendar, Kanban board, forms, or a gallery.

GROUP PROJECTS

Larger projects involving teams need more powerful software. Gantt chart apps like MS Project or Smartsheet are useful for building timelines for tasks that are interdependent. Apps like Basecamp, Flow, or Jira are great for assigning tasks and team communication (integral to any good project workflow). Email threads can get long and confusing. In an app like Flow, all team communications are organized and easy for team members to follow.

There are even enterprise level tools out there for corporations and government agencies. NASA developed its own, called Program Management Tool (PMT), that it uses across all its organizations.

FOLLOW-THROUGH

After everything's done, project management doesn't end. A good postmortem can yield a lot of knowledge that can help you on the next project.

Whatever system you chose to track projects, it has to be a system that everyone buys into and fits the way you or your team works. Try out a couple different options, and see what clicks.

When planning, be realistic about the amount of time things take. Sure, you can work 16-hour days, but you will be exhausted, make mistakes, and create resentment within the team. Also, while working, always keep your timeline up to date. If a task took two days longer than planned, and you have a concrete deadline, you need to decide where to make up that time.

I've worked in project management for years, but I've always been a frustrated artist and maker on the side. When I finally applied some of the tools I use at work on my own projects, I actually got things done, and that made things way more rewarding. You can too.

JAKE GUTTORMSSON is a project manager, artist, beekeeper, and hot sauce maker who lives with his family in N.J.

Work It!

HOW MAKERS ARE HACKING THE 40-HOUR WEEK

WRITTEN BY DANIELLE ZIMMERMAN, VP QUICK BASE BUILDER COMMUNITY

The tech world needs creative risk takers like makers. A lot of companies don't know how to develop the mindset of "I'll try it" in their people. That's what innovation is really. All the maker movement is doing is helping people develop their own sense of experimentation. —Dale Dougherty, founder of *Make:* magazine and Maker Faire

There couldn't be a better time to bring your **whole self** to work. Those awesome things that make you a "maker" also make you a great problem solver, and the good news is you don't have to leave that part of you at home. There is nothing more satisfying than making and building something yourself. You know the sensation … you feel like a badass! But it's not always easy to find the time to incorporate all of those projects you want to do. Here are some work-life hacks that can help you find time to make and build.

1. IDENTIFY YOUR TIME-SUCKS

As busy as we are, not all of that time is productive, even enjoyable; there are a ton of distractions at home and at work. Jot down how you spend each day for a week; if you want to get fancy, you can even create a pie chart that will help you visualize the types of things you spend your time doing. This will also help you identify the things you could probably swap out for personal pursuits.

2. RESERVE 10% OF YOUR TIME FOR "WHITESPACE PROJECTS"

There are many studies that show blocking a certain portion of your time at work for unstructured "play" and experimentation can not only boost happiness and get you exercising your natural curiosity and creativity, but it can also boost engagement and productivity as well. Grab some co-conspirators and make it a team-building exercise, find a nice space and collaborate on a project together!

3. FOCUS ON ONE PROJECT AT A TIME

Have trouble starting a bunch of DIY projects and never *quite* finishing? Create a Kanban board (Japanese for "signboard" — made famous by an industrial engineer at Toyota) to focus on one project at a time and visually see your progress as you go. They can also be fun to make — the more creative, the better. At the top of your board, create three columns: To Do, In Progress, and Done. Write out all the tasks you'll need to do to complete the project. Then move those tasks across the board as you work. Remember you can only work on one thing at a time!

Go be you and keep on building!

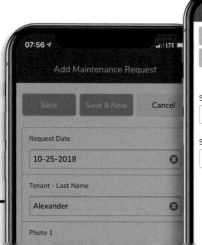
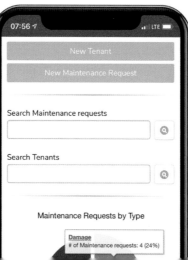
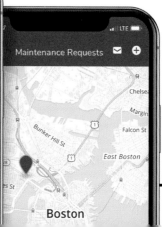

Build your own 3D-printed large-format camera to make incredible images

Written by Drew Nikonowicz

DREW NIKONOWICZ is an artist from Saint Louis, Missouri. He uses new and old technologies to make cameras and photographs and just generally tinker.

The Big Picture

The Standard 4×5 is a large format analog view camera. It uses film to capture images like a 35mm camera does — however, rather than a long roll of film, this camera uses individual 4×5 inch sheets to create photos that are sharper, richer, and 13 times bigger. I designed and 3D-printed my first working prototype in 2014 (Figure Ⓐ), then shared it with colleagues who helped me improve it. Now I'm ready to share it with everyone.

Large format photography is a traditional form of image-making, and view cameras are powerful yet simple. The lens projects the image (upside down and backward) onto a ground glass screen (Figure Ⓑ) that lets you view exactly what photo you'll get when you replace the glass with the film sheet. And the flexible bellows lets you move the lens (and even the film) in different directions to change the plane of focus and the effects of perspective. You can learn the fundamentals of photography by creating and using a view camera.

And once you've built it, the Standard is a legitimate option for anyone interested in large format. It's the first camera I reach for when shooting 4×5. The front and rear standards have a wide range of movements and they accept Linhof lensboards and Graflok backs. The modular design makes it easy to replace parts or modify the camera. It even has a bubble level for perfect horizons.

If you'd like to make the Standard 4×5 camera from scratch, visit manual. standardcameras.com to download the 3D print files, cutting templates, and bill of materials (Figure Ⓒ). You'll also find thorough instructions and resources for construction and assembly.

Photographers of all ages will enjoy building The Standard 4×5. Using it feels like using a piece of history. Finishing this project will leave you with a functional camera that can make incredible images and help you develop a better understanding of photography. I hope that by making an open source camera, I can help more people discover and enjoy large format.

BUILD YOUR STANDARD 4×5 VIEW CAMERA

Thorough construction details and variants can be found on the Standard Cameras website; here's an overview of how it all goes together.

The first working prototype, 2014.

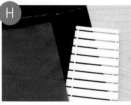

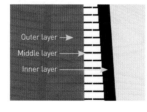

Drew Nikonowicz, India Ivy

TIME REQUIRED:
3–4 Days

DIFFICULTY:
Intermediate

COST:
$100–$150

MATERIALS
» **3D-printed parts (41)** PETG filament is recommended. Download the STL files for printing at manual.standardcameras.com.
» **Aluminum square tube, 1"×12" long**
» **Aluminum bars, 3/8"×1/8":** 7¼" (2) and 8⅛ " (2)
» **Glass, 1/8" ×4"×5" (2)**
» **Camera bellows materials** see Step 4
» **Bubble level, 10mm×6mm**
» **Adhesive rubber pads, 34mm×55mm (2)** for grip between forks and rail slider
» **Felt, adhesive** for light-tight lensboard seal
» **Elastic cord**
» **¼-20 tee nut**
» **Assorted screws, nuts, and washers** See the BOM online for a full list.

TOOLS
» **3D printer**
» **Band saw or hand saw**
» **Hand drill, drill press, or milling machine**
» **Silicon carbide powder, 600 grit**
» **Scissors or rotary cutter**
» **Phillips screwdriver**
» **Utility knife**

1. 3D print the parts
Before you begin, make sure the filament you're using is light-tight, and see how many layers are needed to achieve this. There's a 3D file included for testing this.

I recommend black PETG for the camera parts (Figure D). However, any part that doesn't face the inside of the camera can be printed in any color you like. Every part has been designed so that support material should not be required. If this seems impossible, try changing the orientation of the part in your slicer.

2. Make the aluminum parts
There are 5 aluminum parts in total: a 1" square rail, and four 3/8"×1/8" bars for the adjustable fork pieces.

Cut the square rail to 12" and debur (Figure E).

Included in the download packet are STL files for the four aluminum fork pieces. There are two versions of the front forks.

Version 1 is for making the part with a milling machine. If you don't have a mill, Version 2 requires only a drill — ideally a drill press. (I don't recommend it, but these parts could alternatively be 3D printed in a pinch. If you do this, use a rigid filament like PLA with a high infill percentage.)

3. Make the ground glass
First cut down the glass to 4"×5". Many hardware stores will do this for free if you can't do it at home. For a proper fit, the glass shouldn't be thicker than 1/8". Make at least two pieces.

Use one of the pieces to create a tool like the one in Figure F: super-glue a scrap of wood onto the glass to make a handle.

Tape one side of the ground glass to a piece of cardboard so it can't move while you're grinding it. On top of the glass, make a 1:1 slurry of 600 grit silicon carbide powder and water. Using the tool you made, apply moderate force and make circular

motions to grind the glass for at least 5 minutes (Figure G). Grind the entire surface of the glass until it has a translucent "frosted" look, and only do this on one side! Rinse the glass and let it dry — if there are still some areas where the glass is smooth, repeat the grinding process.

You can add grid lines to aid focusing and composition, by lightly drawing on the ground side with a sharp pencil and ruler.

4. Make the bellows
This is the most complicated part of the process. Follow the guide on our website.

There are three main components to a bellows: an inner layer of light-tight material, a middle layer of ribs or stiffeners, and an outer layer of protective material (Figure H). It is essential that the bellows is light tight, and as thin as possible. Be sure to check before committing to a material. Many people use thin leather or vinyl for the inner material, but the best is a coated nylon. The stiffeners can be cut from a manila folder or something of similar thickness. The outer material can be anything — ideally something thin and strong like ripstop nylon.

To glue everything together, use a spray contact adhesive for the best results.

5. Assemble the camera

Finally for the fun part! All you need is a screwdriver and utility knife to complete the camera (Figures ⓘ and ⓙ).

You'll assemble the rail with its tripod mount and standard mounts, install the pinhole lens in the front standard and the ground glass in the rear, then connect both standards to the bellows. Finally, you'll attach the adjustable forks to the rail, and the standards to the forks. Follow the manual online for detailed step-by-step instructions. Be careful not to overtighten the screws that thread into printed parts.

Then double-check that everything works properly. Check the pinhole lens image on the ground glass, rotate the rear standard from landscape to portrait mode, and test all the standard movements (see "Using a View Camera" on the following page).

GET SNAPPING

Congrats, you have successfully made your very own Standard 4×5! Now you just need a few things and you're ready to head out the door to start shooting.

» First you'll need a **lens**. The 3D-printable pinhole lens included in the download packet is a good starter lens. If you want something more powerful, just be sure it is attached to a Linhof or Wista style lensboard.
» You will also need a sturdy **tripod**.
» For focusing, a **"dark cloth"** is needed to properly see the ground glass. However, an extra raincoat or sweatshirt works just fine and if it rains you're prepared!
» Finally, 4×5 **film holders** are needed to properly load your film into the camera. Most brands should fit, so don't worry about buying a specific one.
» You might also want a **light meter** to determine your camera settings, but you can also use a digital camera's meter if you have one.
» You're ready to start shooting! Buy some **4×5 sheet film** and get out there.

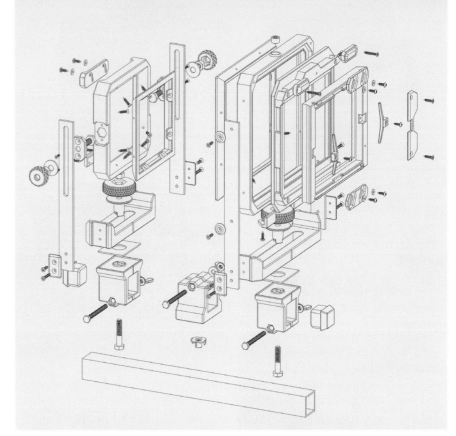

The Standard 4×5 camera 'in the field' by photographer India Ivy.

Camera in use by photographer Asa Lory at an event at the Saint Louis Art Museum.

Camera in use by photographer Asa Lory at an event at the Saint Louis Art Museum.

Camera in use by photographer Marissa Dembkoski.

USING A VIEW CAMERA

With a flexible bellows, view cameras have extra functions that ordinary cameras don't — and you'll see the effects instantly on your ground glass. Here's what the standard movements do:

Rise/fall — Vertically shift the front standard higher or lower to change the portion of the lens image that's captured on the film, or to straighten convergent lines.

Shift — Same idea, but shift the standard(s) horizontally.

Tilt — Tip the front standard forward or back, to chart a plane of focus through your image and find the best depth of field for your subjects.

Swing — Turn the standard(s) left or right, to swing the plane of focus sideways. In Figure K you can see the front standard *tilted* forward and *swung* left. The rear standard can also be swung to adjust focus and to correct or exaggerate perspective.

I would love to see your Standard 4×5 cameras and the photos you make with them. I hope everyone will share them with me on Instagram at @standardcameras. The photography community was a big motivation for this project, so I'm very excited to see what everyone makes! ✪

Living Large

Here are some photos shot by the Standard 4×5 community using their DIY large-format view cameras.

Drew Nikonowicz (wormhole, scaffold), Asa Lory (crouch, curtains), Joe Johnson (type blocks, tree)

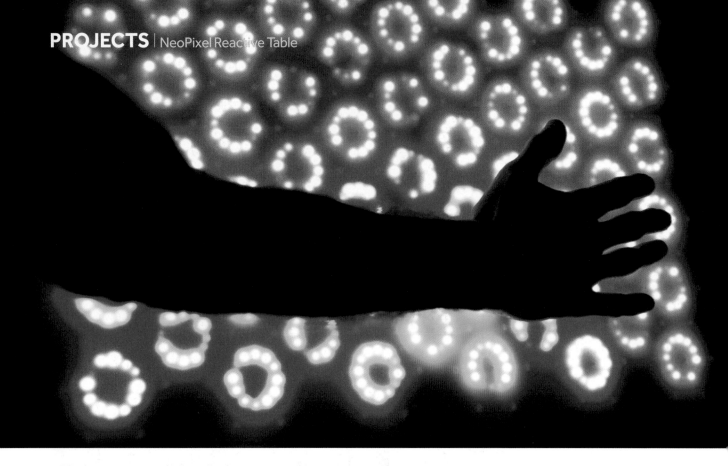

The Playful Table

Deploy a fleet of smart LED rings and simple sensors that react to motion with blazing, animated light displays   **Written by Sam Guyer**

A couple years ago I saw a YouTube video of a reactive LED coffee table that blew my mind. Under the glass tabletop was a grid of blue LEDs that lit up in response to anything placed on or near the table. Sensing was implemented using an array of infrared emitters and photodiodes scattered among the LEDs. Invisible infrared light from the emitters reflected off ordinary objects and was detected by the photodiodes. The demo was mesmerizing, and I couldn't wait to start building my own.

When I looked at the complexity of the design, though, I got discouraged. In particular, wiring and controlling a large number of discrete LEDs is a daunting task. Then it hit me: Why not use addressable LED strips, such as the ones based on WS2812 smart LEDs? Using this strategy, a single data wire can control hundreds or even thousands of individual LEDs. As an added bonus, they support full 24-bit color, allowing a practically unlimited range of reactive patterns.

My design uses rings of LEDs rather than strips, each surrounded by six IR emitters with a single IR sensor in the middle. Together, these components form a reactive "cell": the software continuously reads the value from the IR sensor and renders some pattern on the LEDs in response. The pattern can be as simple as turning on individual LEDs or changing their color, but the most fun patterns are animated ones that flash, spin, or sparkle.

Here's how to build a single cell and an overview of how to chain multiple cells together. The complete table shown here consists of 61 cells, which is a substantial amount of work. But even a single cell is fun to play with! In the process you'll learn how to wire and read IR sensors and how to program addressable LED strips.

1. DRAW THE LAYOUT

The LED and IR components are mounted on cardboard or foamcore to hold them in place. In my design, the cells are organized as a hexagonal grid (Figure Ⓐ), with one IR emitter at each of the six vertices. The IR photodiode sits at the center surrounded by the LED ring.

2. PUNCH HOLES FOR IR COMPONENTS

The electronic components sit on the top surface of the cardboard, but all of the wiring is hidden on the back. So we start by punching holes through the cardboard in a pattern that matches the leads of the components.

To ensure proper spacing for the IR emitters and sensors, I built a hole-punching "jig" by soldering a two-pin piece of right-angle header, which exactly matches the LED leads spacing, onto a small protoboard (Figure Ⓑ). You can add some hot glue and sharpen the points to make it easier to punch.

Start by punching holes for the emitters at each corner of the hexagon. Next punch a pair of holes for the emitter in the center.

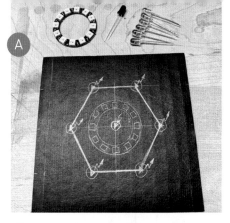

3. PUNCH HOLES FOR RGB RING

Each RGB LED ring has six pads on the back: three are inputs and three are outputs, which makes them easy to chain together. My strategy for attaching these rings to the board is to solder a three-pin piece of right-angle header to each set of pads, then push these pins through the cardboard, and connect them on the back with three-wire cable.

To punch the holes for the ring I made another jig, which is just a ring with the pins soldered on and sharpened. You can add hot glue to make it stronger. I also glued on two pieces of thin wire to provide a "crosshair" for aligning the ring (Figure C). This is a big help if you're building a bunch of cells.

Line up the crosshair with the center of the cell (Figure D) and push the pins through to create the six-hole pattern (Figure E).

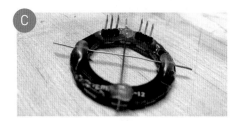

4. INSERT AND WIRE SIX IR EMITTERS

Insert one IR emitter at each corner of the hexagon, with its positive lead on the left, and its ground on the right. On the back of the cardboard (Figure F), fold the leads to hold them in place. Your goal is to connect them into two chains of three emitters in series because these IR LEDs are typically rated for 1.5V or less — too low for our 5V power supply. But three in series will take about 4.5V, and with a small resistor (somewhere in the 12- to 27-ohm range) in series we can power them with 5V. You can see the circuit in Figure N on page 50.

Fold the leads so that the positive lead on one IR emitter touches the negative lead on the next one. Twist them together using pliers. You can add a little solder at the joint. With three connected together, you should have one positive lead at one end, and one negative lead at the other (Figure G).

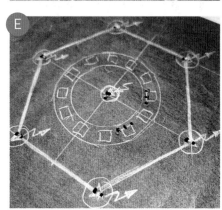

5. ADD POWER RAILS AND RESISTORS

For a single cell, you could just solder leads onto each 3-LED circuit. But if you want to connect a bunch of these cells, add power rails to connect all the circuits more easily. I use a heavier copper wire that you can solder leads directly onto. Add four rails altogether — two power and two ground.

Connect one end of each 3-LED circuit to the nearest rail, and add the small value

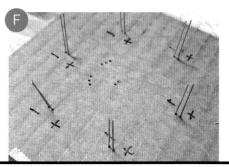

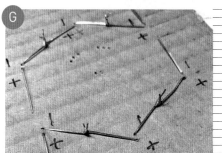

TIME REQUIRED:
1–2 Weeks

DIFFICULTY:
Intermediate

COST:
$20–$30 Per Cell

MATERIALS

» **Microcontroller, Arduino compatible** I chose the Espressif ESP32. It's a great chip: dual core, 160MHz or 240MHz, with plenty of memory, pins, and special features. I worked with a few other people to develop ESP32 support for the FastLED library, in part to make this project possible.
» **5V power supply** See Step 11 for details.
» **Protoboard** e.g., Amazon #B071R3BFNL. Or you could use a breadboard.
» **Analog multiplexer board(s) (optional)** Amazon #B01DLHKLNE, if you're scaling this project up; see Step 11.

FOR EACH CELL:
» **RGB LED ring, 12 pixels, WS2812 type** such as Adafruit's NeoPixel Ring #1643, adafruit.com. For large quantities (and long waits) try aliexpress.com #32673184275.
» **Infrared (IR) LEDs, 940nm (6)** such as Amazon #B01BVGIZGC
» **IR photodiode, 940nm** Amazon #B075B448VF
» **Resistors: 12Ω–27Ω (1) and 10kΩ (1)**
» **Hookup wire, 24 or 26 AWG**
» **Bare copper wire, 22 AWG**
» **Right-angle breakaway header, 6 pins per cell** such as Adafruit #1540
» **Three-wire cable with female Dupont connectors** Amazon #B01N3KGISV; yes, they're confusingly called "male to male"
» **Cardboard or foamcore board**

TOOLS

» **Soldering iron and solder**
» **Pliers, wire cutters/strippers**
» **Computer with Arduino IDE** free download at arduino.cc/downloads
» **Hot glue gun (optional)**
» **Multimeter (optional)**

SAM GUYER teaches computer science at Tufts University. At heart he's an artist and craftsman who loves to build things out of wood, light, and code.

ASG Photography, Sam Guyer

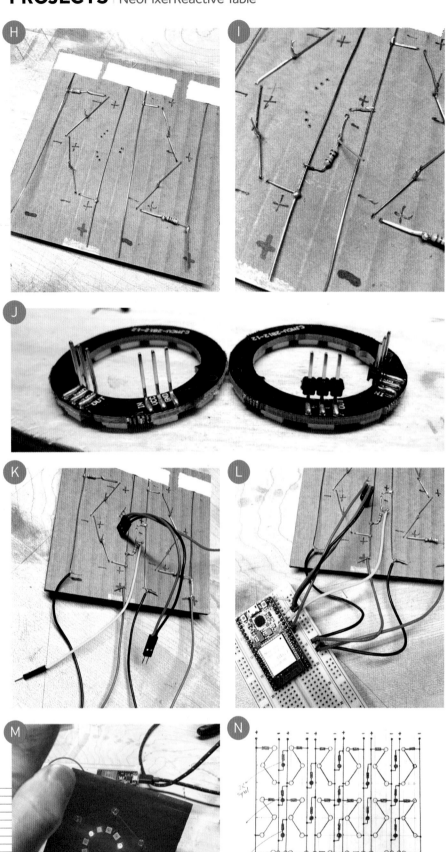

resistor in series on the other end as shown in Figure H.

6. INSERT AND WIRE IR SENSOR

The IR sensor is a photodiode that works in an unusual way called reverse bias: the positive lead (the longer one) is connected to the negative power rail; the negative lead is connected to the positive power rail. As you might expect, this blocks the current — unless there is infrared light shining on it, in which case it allows the current to flow (in the opposite direction from a regular LED!).

It's important to notice that we read the signal on the positive side, which means that the pin will read high (5V) when there is no IR light shining on it (no object in front of it). We add a 10K resistor between the 5V power and the signal to serve as a pull-up and to protect the circuit. When IR light shines on the photodiode, it loses its resistance, and all the current flows directly from power to ground, causing the signal pin to read low (close to 0V).

Insert the IR sensor and fold the positive lead so that it crosses the negative power rail. Fold the negative lead straight down, then add the 10K resistor to connect the negative lead to the positive power rail (Figure I). Leave a little segment of the lead sticking out — this is where you'll attach the wire to read the signal. Crimp and trim all the wires.

7. ADD THE LED RING

Solder two three-pin segments of right-angle header onto the LED ring pads. I suggest you add a dab of solder to two of the pads and "glue" the segments on just using those two pads. Once secure, go back and complete the solder joints on all six pads (Figure J). Carefully use tweezers to remove the black plastic holding the pins together.

Push the 6 pins through the holes. Flip the board over — you should see the 6 pins poking through, with plenty of extra lead. Remember the arrangement of the pads: the ground pins are always the outside pins, the power pins are the middle, and the signal pins are the inside. This setup is important when you want to connect multiple cells.

8. ADD LEADS

Add wires to the two pairs of power rails. I used red for power and black for ground. I

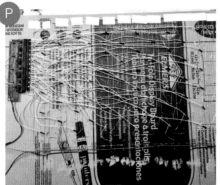

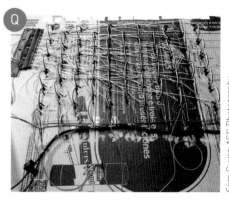

Sam Guyer, ASG Photography

also attached a two-pin Dupont connector to the other end to make it easy to plug in.

Make a three-pin Dupont connector for the LED ring with three wires: power, ground, and data (Figure K). If you want, you can also connect the other end of the power and ground together in a two-pin Dupont connector, just like the power rails. Solder a wire to the output of the IR sensor (the yellow wire in Figure K).

9. CONNECT IT UP!

Solder the microcontroller onto the protoboard (or just plug it in if you're using a solderless breadboard).

Connect the 5V and ground pins on the microcontroller to the appropriate power rails on the protoboard. Connect the three power and ground pairs (red and black wires) to the power rails on the protoboard. Connect the data wire from the LED ring to one of the digital pins on the microcontroller (I chose pin 26 on my ESP32). Connect the IR sensor wire to one of the analog pins (I chose pin 27). Your first cell is complete (Figure L).

10. CONFIGURE AND INSTALL SOFTWARE

Download the ReactTable code from github. com/samguyer/ReactTable, open it in the Arduino IDE, and configure it for your setup. All of the major options are at the top of the source file; in particular, you'll want to define the data pin for the LED ring and the

analog pin for the IR sensor. You can also choose which of the predefined display patterns you want — or write your own! Upload the code to the microcontroller and test it (Figure M).

> **TIP:** How will you know if the IR emitters are working? Most point-and-shoot cameras can see IR light, which shows up as purple in photos. Point your camera at the IR emitters and look at the live image — you should see bright purple light. If you see no light, check your wiring using a multimeter. If the light looks dim, replace the series resistors with smaller ones (lower ohm value).

11. MAKE IT BIGGER

Making a bigger table is mostly about making more cells, but there are a few gotchas to look out for. First, I strongly recommend laying out all your wiring before you start (Figure N). Make sure you have room for all the power rails and the various connections (Figure O).

Second, you will need a way to read a large number of analog inputs (Figure P). The easiest solution is to connect all of the IR sensors to analog multiplexers. I found cheap 16-way multiplexers (based on the CD74HC4067 chip) that connect to the microcontroller with only five wires: four to specify which channel to read, and one for the input value.

Finally, you should be aware that a large table draws a lot of power (Figure Q). Sixty rings with 12 LEDs each can pull as

much as 30 or 40 amps at full brightness, although in typical use it's more like 5 to 10 amps (at 5V). Plus, the IR emitters use some power as well. Buy a 5V power supply that can provide more than the minimum! You can learn more about scaling up this project at instructables. com/id/NeoPixel-Reactive-Table.

PSYCHEDELIC REACTIONS

Now try out your Reactive Table! My software implements four animated LED patterns — Solid colors (Figure R and page 48), sparkling Confetti, rotating Gears, and flickering Fire (Figures S, T, and U), which you can see in the demo video on the Instructables page.

Most of the interesting stuff happens in the **Cell** class; there's one instance of this class for each cell you build. Each pattern is implemented by a method in the class that reads the IR sensor and sets the colors of the LED ring — only for that cell. The main loop of the sketch just calls this method repeatedly at the given frame rate.

You can learn more about the code at the Instructable and by reading the code itself; it's thoroughly commented to show you what's happening each step of the way.

There are so many directions you can take this project. Add more cells, or organize them in a different way, or design your own reactive animations. It's a blank canvas just waiting for you to dream up something cool! ◆

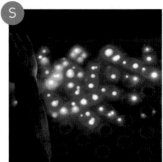

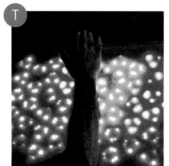

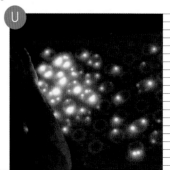

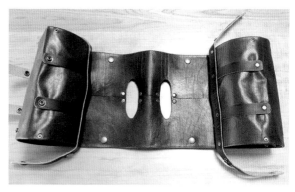

This **leather carrier** straps onto your bike (or broomstick), and holds a six-pack or two wine bottles

Convertible Caddy

Written and photographed
by Mikaela Holmes

MIKAELA HOLMES is a costume and experimental fashion designer, artist, and writer whose work fuses couture sewing and leatherwork with wearable tech. mikaelaholmes.com and floraandfaun.com

A

Be the classiest maker at your next picnic with this leather carrier that straps onto your bike (or broomstick), transforms to carry beer or wine, and collapses into a neat leather scroll when you're done!

There are a lot of six-pack carriers out there, but I wanted something multi-functional that could be disassembled and folded into a small package. I experimented with different ways of folding, prototyped it in paper (Figure Ⓐ), and managed to come up with a design that securely carries six beers or two bottles of wine, can be easily attached to a bicycle and comfortably carried by hand, completely disassembles, and rolls into a little scroll that snaps closed. The construction requires no sewing, and the pattern is just a simple rectangle and four straps.

I'm a huge *Harry Potter* nerd, so I couldn't resist stamping "I solemnly swear that I am up to no good" onto the straps. I'd like to think Moony, Wormtail, Padfoot, and Prongs would approve. (But they wouldn't approve of drunken bike riding, so be responsible.)

Only muggles carry beer in cardboard six-packs.

1. TRACE THE PATTERN

Download the pattern (Figure Ⓑ) from my project page at instructables.com/id/Leather-Beer-Wine-Carrier and print it out full size.

Spread out your leather, underside facing up. Place the "Unfolded" pattern on top and secure it with small pieces of tape. Outline the edges in pencil, then use the sharp edge of an awl to trace all the internal fold lines

B

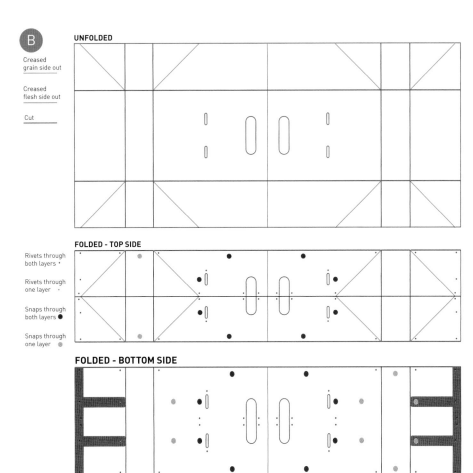

UNFOLDED

Creased
grain side out

Creased
flesh side out

Cut

FOLDED - TOP SIDE

Rivets through
both layers ·

Rivets through
one layer ·

Snaps through
both layers ●

Snaps through
one layer ●

FOLDED - BOTTOM SIDE

through onto the leather. This will tear the
paper a bit.

2. SCORE AND CUT

Use the adjustable V gouge to score the
internal fold lines (marked in green on
the pattern) straight and even (Figure C).
Make sure the gouge is set to a very shallow
depth, and test it on a scrap first — this
leather is fairly flexible and doesn't need
much gouging to fold well. A little goes a
long way.

Then use sharp scissors to cut out the
rectangular outline of the pattern that you
marked with pencil.

3. FOLD

Fold the leather along your gouged lines
and pound the folded edges down with
a hammer (Figure D) — first the inner
creases that create the two pockets, then
the two sides that fold to meet in the
center. Check the pattern to see which
direction to fold. Don't pound too hard or
you'll leave marks — you can prevent this
by putting another piece of leather over
the fold and pounding through that.

4. GLUE

Those two main flaps now need to be glued
down everywhere except where the pocket

C

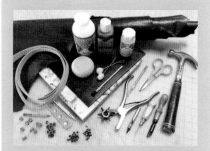

D

E

is going to open. Check the pattern and
mark off the areas to be glued, then use a
disposable brush to cover the areas in glue
(Figure E). Apply leather contact cement
to both surfaces, wait 10 minutes until it's
tacky, then press the flaps down, making
sure they meet in the middle and lay flat

TIME REQUIRED:
6-8 Hours

DIFFICULTY:
Intermediate

COST:
$75-$150

MATERIALS

» **Medium weight leather (3oz–4oz),
vegetable-tanned, about 34"×16"**
for the body of the carrier, such as Tandy
Leather #9399-99 or 9048-66 (or 9828-
97 if you make sure to request a thicker
piece), tandyleather.com
» **Thicker leather straps (4oz–5oz),
¾" wide: 12" long (4) and 19" long (2)**
I bought two 50 " veg-tanned pre-cut
straps, Tandy #4523-00.
» **Leather stain, water-based** I used Eco-
Flo Waterstain in black, Tandy #2800-01.
» **Leather finish** to finish the water stain. I
used Eco-Flo Satin Shene, Tandy #611-01.
» **Snaps (20) with setter** Tandy #1261-04
or similar
» **Rivets, quick-set (36)** Tandy #1271-15 or
similar; rivet setter optional
» **Leather contact cement** Tandy #2525-01

TOOLS

» **Sponges or wool daubers** such as Tandy
#3445-00, for applying stain and finish
» **Hole punch, manual** Tandy #3777-167
» **Rotary punch (optional)** Tandy #3240-
00; it's easier to use for some of the holes,
but you still need the manual punch for
holes a rotary punch can't reach.
» **V gouge, adjustable** Tandy #31811-00
» **Awl** Tandy #3209-00
» **X-Acto knife**
» **Hard surface** for setting rivets,
hammering stamps, etc. I use a quartz
slab, Tandy #32228-00, with a hard
rubber "poundo" board on top, Tandy
#3461-151, but you can also use a concrete
floor, granite countertop, or piece of steel,
or a wooden workbench if you've got a
metal rivet/snap setter base.
» **Hammer**
» **Leather stamps (optional)** to decorate
your straps. I used an alphabet set, Tandy
#4903-01.
» **Scissors** for paper and leather
» **Ruler and pencil**
» **Chalk**
» **Tape**
» **Large piece of paper**
» **Wax paper or plastic wrap**
» **Disposable brush**

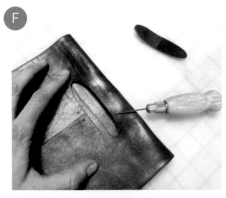

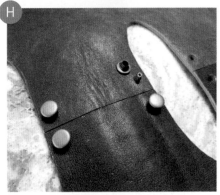

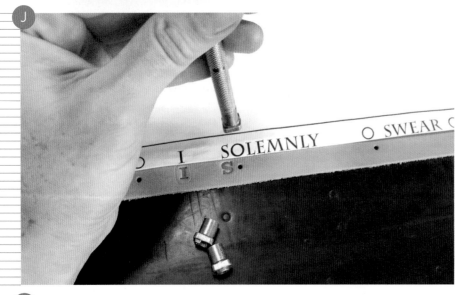

with no wrinkles. I used my hammer on the folds again to make an extra sharp crease.

5. PUNCH

Use the pattern to trace the handle holes and strap slits onto the leather (Figure F), then cut them out with a sharp X-Acto knife. I also used a hole punch to define the ends of the slits and make them easier to cut out.

Then measure, mark, and punch each rivet and snap hole (Figure G).

6. RIVET

Pound in the rivets using a hammer, with their smooth cap end facing toward the side of the leather where the folded edges meet, because this is the side that will mostly be visible (Figure H). I used quick-set rivets, which don't need any special tools to attach them — very convenient and reliable for small projects like this.

7. MAKE STRAPS

Use scissors to round off one end of each short strap, then mark and punch holes ½" in from the rounded ends (Figure I).

Punch 4 holes in the carrier body where the 2 long straps will attach, then use these holes to mark the long straps and punch them as well.

8. STAMP STRAPS (OPTIONAL)

Before I added my message to the straps, I made a mock-up in Illustrator and played around with word placement. I decided I needed to use two of the rivets as letter O's in the words in order to fit them all, and I like the way it looked in the end.

Before stamping, wet (or "case") the leather with a sponge to soften it. Then use a hammer to pound your stamps into the leather (Figure J). It was tricky to keep them in a straight line, even using my printed mock-up. I even stamped a couple of them upside down by accident, which is easy to do because you can't see the orientation of the letters from the backside. I was able to smooth over these mistakes and re-stamp, but the results weren't quite as neat as I had hoped. I just had to embrace the glitch because I didn't have any more straps.

9. DYE STRAPS

To dye your straps, lay them out on wax paper and use a wool dauber or sponge to apply water-based stain. I decided to leave the edges raw, so I only applied dye to the

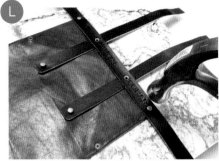

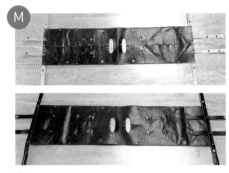

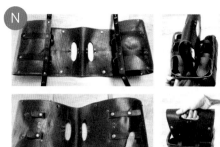

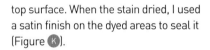

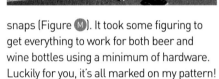

top surface. When the stain dried, I used a satin finish on the dyed areas to seal it (Figure K).

10. ATTACH STRAPS

Use the straps to mark the remaining rivet holes on the carrier body, then punch holes. The bottom ends of the two short straps attach through just one layer of the carrier, allowing the other layer to fold up and create a pocket, so take care not to punch through both layers. I slid the corner of my poundo board between the two layers to use as a punching surface.

Now rivet the black straps onto the carrier (Figure L). The short straps go under the long straps and the two are riveted down together. (I strategically inverted the rivets in the center of my long straps where I used the rivet heads as part of the lettering.)

To safely rivet the bottom ends of the short straps without damaging the second layer of folded leather, I slid the end of a metal ruler between the layers of leather and pounded the rivets onto that. On the front of the carrier, you'll use male snaps for these bottom ends, instead of rivets — for securing the straps when the whole case is rolled up.

11. SET SNAPS

When all straps are attached, mark where the two sides of all the snaps should go on the straps and carrier body, then set the

snaps (Figure M). It took some figuring to get everything to work for both beer and wine bottles using a minimum of hardware. Luckily for you, it's all marked on my pattern!

» For the short straps, I added two sets of snaps on the carrier that allow them to either attach for beer carrying, or cinch down tighter to wrap around a wine bottle on each side.

» The long straps snap around onto each other to secure the two sides of the beer carrier, or snap down over the ends of the bottles when the carrier is in wine mode.

» To hold the neck of the wine bottles, I made slits in the long straps on one end large enough for the neck to slip through.

» I added two snaps that hold together the two sides of the carrier near the handle, which helps it stay secure on a bike.

» To allow the whole thing to roll up and snap onto itself, I added a second set of snaps to the top ends of the short front straps.

TRANSFIGURATION

Now your beverage carrier can reveal its magical shape-shifting capabilities!

BEER CARRYING MODE: Slip the straps through the slits and snap them onto the upper snaps, leaving a gap between the layers of brown leather where the beer will go. Then reach in and pull out the inner layers of the leather so they pop up and form boxes that will hold the beer.

Snap the horizontal straps around the ends of the carrier and close the two snaps near the handle to hold the whole thing together (Figure N). If you're attaching it to your bike, wait to do these last two things until you have placed the carrier over the bar of the bike.

WINE CARRYING MODE: Pull the short straps through the same slits as far in as they'll go, and attach them to the lower snaps. This creates a tube on each side that will hold the wine. The horizontal straps on one end have slits in them; snap these down to the snaps on the bottom of the leather tubes. To secure the wine, slide the wine bottles into the tubes and stick necks through the slits, then snap the second set of horizontal straps around the other ends of the bottles (Figure O).

Attach the carrier to your bike the same way (Figure P).

STORAGE MODE: Roll up the carrier, snap it closed with the short straps, and tuck the long straps inside (Figure Q).

[+] See more photos and tips at instructables.com/id/Leather-Beer-Wine-Carrier.

Written and photographed by Nathaniel Bell

TIME REQUIRED:
About 4 Hours

DIFFICULTY:
Easy

COST:
$20–$40

MATERIALS

» **Wood, acrylic, or other sheet material: ¼" or ⅛" thick, 6"×23" (2)** You can use multiple thicknesses, just note which parts must be the same thickness in the steps. I used walnut hardwood. Thin stock craft wood packs from either Klingspor or Woodcraft work well, or you can find them online at Cormark International, cormarkint.com.
» **Wood dowel, ¼" diameter, about 12" total length**
» **Wood screws, small (4)**
» **Wood glue**
» **Cyanoacrylate (CA) glue** aka super glue
» **Handle** of your choice. A thicker dowel or rounded stock works well.

TOOLS

» **CNC router (⅛" mill) or laser cutter**
» **Clamps (2 or more)** Eight would be great.
» **Screwdriver**
» **Handsaw to cut dowel**
» **Drill and ¼" bit**
» **Sandpaper**

Switch Gears

Build this fun and totally unnecessary mechanism to flip your lights off and on

NATHANIEL BELL is lead artist at Insomniac Games NC, a father of three, chicken aficionado, and maker of things in Hillsborough, North Carolina. Follow him at instagram.com/stuffnatemakes.

I was initially inspired to make a gear-driven light switch when I saw one that a friend of mine bought online, which I later learned was inspired by a "light switch complicator" that Ernie Fosselius showed at Maker Faire. You turn the crank, and a series of gears translates a linear slider that flips the switch.

I found this to be a relatively easy project to design. It looks complicated, but with the help of gear creator sites like geargenerator.com, the hardest part was creating a layout to keep everything compact (Figure A). You can cut the files on a CNC or a laser cutter

in hardwood or whatever materials you think look nice. Here's my version in solid walnut. It's a fun project that can be done in an afternoon.

1. CUT THE PIECES

Download the plans (makezine.com/go/light-switch-complicator) and cut all pieces out with either a CNC or laser cutter for precision (Figure B). All screw holes should be countersunk, and functional pockets should be about half the depth of the material they are in.

Gears, spacers, and the switch flipper all

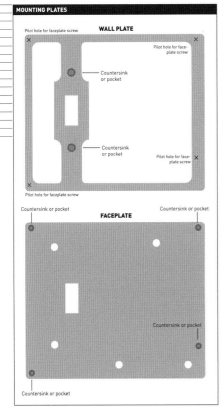

MOUNTING PLATES

WALL PLATE

Pilot hole for faceplate screw

Pilot hole for face-plate screw

Countersink or pocket

Countersink or pocket

Pilot hole for face-plate screw

Pilot hole for faceplate screw

Countersink or pocket

FACEPLATE

Countersink or pocket

Countersink or pocket

Countersink or pocket

Countersink or pocket

FLIPPER

FLIPPER TRACK

Pocket (decorative) ½ depth

Pocket (functional) ½ depth

Pocket (decorative) ½ depth

SWITCH EXTENDER

Adhere to side of light switch to make switch reach the flipper track.

Adjust length based on thickness of material

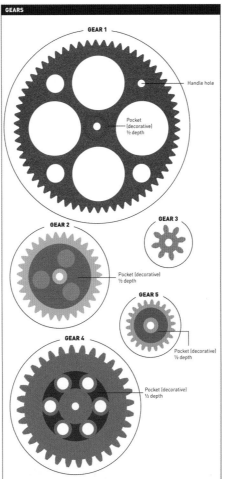

GEARS

GEAR 1

Handle hole

Pocket (decorative) ½ depth

GEAR 2

Pocket (decorative) ½ depth

GEAR 3

GEAR 5

Pocket (decorative) ½ depth

GEAR 4

Pocket (decorative) ½ depth

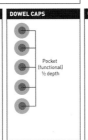

SPACERS

Used on top of flipper track. Rides in the functional pocket to keep track aligned.

Used under flipper track to raise it up to the level of gear 5.

1st layer   2nd layer

Used under gear 4 to raise it up to the level of gear 3.

Pocket (functional) ½ depth

Gap for gear 4 clearance

DOWEL CAPS

Pocket (functional) ½ depth

GEAR BACKS

4   1

5   3   2

Used on backside of the face-plate to give gear shafts a thicker surface to attach to.

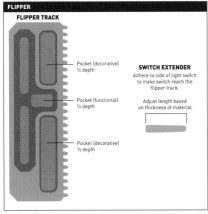

need to be the same thickness (Figure C). The faceplate, wall plate, end caps, switch extender, and track alignment spacers can all be a different thickness if needed. A thickness of ¼" for gears and ⅛" for plates and extender is a good place to start.

IMPORTANT: Take care to choose stock that is as flat as possible or the gears will bind due to the imprecision.

2. MAKE THE SHAFTS

Start by gluing the end caps onto five ¼" dowels, as square as possible (Figure D).

Leave these long: 2" or longer. You'll cut them to the correct length later.

Glue two of the octagon-shaped spacers to the end caps (Figure E). These two will ride in the flipper tracks to keep it aligned.

3. GLUE SPACERS

Glue the gear backs to the back of the faceplate (Figure F). These should be the same thickness or thinner than the wall plate; they're to give the faceplate holes more depth to glue the shafts in later. You can use the wall plate as a guide for placement when gluing.

Glue together and set aside the two sets of spacers that will be used under the flipper track. The two solid spacers go together as a pair. The spacer with the pocket and the one with the matching cutoff go together as shown (Figure G).

4. GLUE THE GEARS

Glue gears 2 and 3 together (Figure H). The tooth alignment does not matter at this point.

Glue together gears 4, 5, and the large round spacer. Use a shaft to align all the center holes to guarantee smooth rotations, but be careful not to glue the gears to the shaft (Figure I). They will need to spin freely on it later.

5. MOUNT THE GEARS

Add the 3 shafts that only have end caps to each of the gear sets (Figure J). Sand the shafts if needed to make the gears spin easily without friction.

Slide the gears into position on the faceplate (Figure K). At this point, turning any of the gears should rotate the whole assemble easily. Out-of-square gear shafts will cause binding.

6. MOUNT THE SWITCH FLIPPER

Glue the two sets of octagon spacers from earlier to the faceplate (Figure L). These are offsets for the flipper track to sit on. Align the lower set so the cutout allows clearance for gear 4 to spin (Figure M).

Inset the two shafts with the octagon track spacers into the flipper track (Figure N). Then insert the flipper track into the stacked spacers (Figure O).

Check alignment. Rotating gear 1 should easily move the track up and down. If there's friction, try sanding the sides of the spacers that ride in the track to create clearance.

7. GLUE THE SHAFTS

Align the gear positions so the smallest (¼") hole on gear 1 is in the upper right corner position when the center hole on the flipper track is centered over the switch hole on the wall plate (Figure P).

Once everything is aligned, mark the depths on each shaft, remove, add wood glue, and reinsert (Figure Q). Double-check alignment before the glue dries.

8. ADD A HANDLE

Drill a ¼" hole and insert a dowel in a piece

GEAR PLAN

of rounded stock (Figure R) to use as a handle. For mine I used a ½"×½" scrap, rounded on a belt sander (Figure S).

Insert and glue the handle into the ¼" hole in gear 1.

9. PREPARE THE LIGHT SWITCH

Remove the wall plate from a single standard light switch.

Using the screws for the original wall plate, attach the new wall plate (Figure T). Make sure the screws are flush or below the surface. If not, countersink the screw holes. The faceplate must sit flat on the wall plate.

Using super glue, attach the switch extender (Figure U). Make sure the switch can still move freely to change positions. Leave it long and cut it to size after attaching the faceplate with gears attached.

10. MOUNT IT ON THE WALL

Attach the faceplate to the wall plate with small wood screws in each of the four corners (Figure V). The lower right screw hole should be accessible through the holes in gear 2 based on the rotation.

Depending on thickness of your wall plate and faceplate, you may need to grind your screws short. Again, make sure they're flush or below the faceplate surface, or the gears will not turn smoothly.

You're done. Now turn the handle to unnecessarily complicate your switch flipping!

GOING FURTHER

You can use the free calculator at Gear Generator (geargenerator.com) to design your own gearing, or to create an all-new light switch complicator.

I've made this twice now, once with my CNC and once with the laser, and both ways worked without any special modification. I thought for sure I'd have to change something for it to work with a laser but I was able to engrave the pockets just fine.

I guess you could 3D-print the pieces as well, but 3D-print stuff looks like junk and why would you want junk on your wall? ◗

[+] Get the free plans and more photos at makezine.com/go/light-switch-complicator.

Threads of Knowledge

Celebrate 150 years of Mendeleev's breakthrough periodic table by tying 200,000 tiny knots **Written by Jane Stewart**

EDITOR'S NOTE: While we were preparing this article, Jane interviewed herself just for fun and we liked it, so here it is.

Since then, she has been invited to show her macramé periodic table at several 2019 events celebrating the 1869 publication of Dmitri Mendeleev's periodic table of the elements, including the International Year of the Periodic Table opening ceremonies in Paris and exhibitions at the University of Edinburgh, St. Catharine's College in Cambridge, and the Royal Society of Chemistry in London.

At press time, she was just polishing off the actinide (aka actinoid) series.

Why the periodic table?
My dad is a retired chemist (and hobbyist cider maker, see makezine.com/projects/kitchen-table-cider-making). After my last project, he suggested I do a knotted version of the periodic table. The table has its 150th anniversary this year so it's very timely.

What materials are you using?
I'm using metallic crochet thread, which is about the same thickness as embroidery thread. I'm a big fan of rainbow colors so I have given a different color to each group in the periodic table in rainbow order.

How it is done?
About 200,000 half hitch knots. Two for every "pixel" of color. Knotting right to left

with color threads and left to right in black to do the writing.

How long?
I timed myself: It takes 5 minutes for one row in a square, so in theory that is 2 hours per element square, working flat out. And there are 120 squares!

What will you do with the project?
Display it at craft fairs or art galleries. Plus I will take it with me in my STEM ambassador role promoting science in U.K. schools (stem.org.uk).

What's next?
Macramé 3D shapes, as toys for my nephews.

TIME REQUIRED:
6 Months

DIFFICULTY:
Intermediate

COST:
$100–$120

MATERIALS
» **Embroidery floss or crochet thread**
  I recommend black for the structural thread, and your choice of colors for the working threads.

TOOLS
» **Scissors**
» **Graph paper and pen** for planning

JANE STEWART is a lifelong crafter. For the last few years she has been tying many, many knots in things. This is the ancient art of macramé.

[+] Get all the Periodic Table patterns and more photos at makezine.com/go/macrame-periodic-table. To get started with macramé, check out makezine.com/projects/macrame-101.

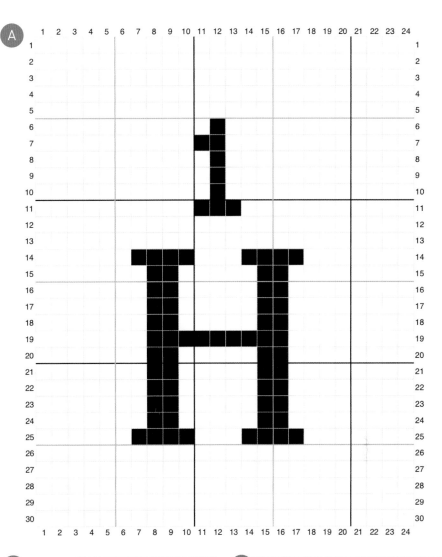

HOW TO MACRAMÉ THE PERIODIC TABLE

The structural thread is black. The working thread is white in the first square (Hydrogen); you'll use different colors for different elements.

Follow the pixel pattern (Figure **A**) on its side, knotting right to left for 30 rows in white. You'll make two knots to create each pixel (Figures **B** and **C**). You can download the patterns from the project page at makezine.com/go/macrame-periodic-table.

When black appears on the pixel pattern, knot the black thread left to right around the white thread so a black pixel will appear (Figure **D**).

The black thread didn't photograph well, so let's take a better look at the two types of knots. In Figure **E**, the working thread (white) is knotting right to left; do it twice.

In Figure **F**, the structural thread (pink) is knotting left to right to cover the main color; do it twice.

Finishing off a pink pixel, you see a single pink dot amongst the white (Figure **G**).

Repeat 200,000 times. ✦

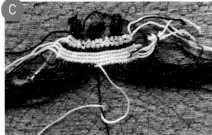

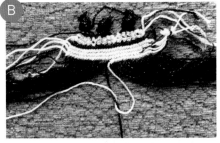

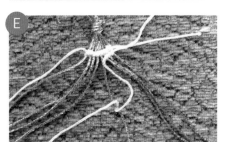

Dr. Helen Stewart

Track Your Stats

Show your social media numbers in real time with seven-segment LED displays and an ESP8266

Written and photographed by Becky Stern

BECKY STERN is community product manager at Instructables and the author of hundreds of tutorials, from microcontrollers to knitting. Previously, she was a video producer for *Make:* and director of wearable electronics at Adafruit.

Smokey Nelson

Frame your followers! Here's how to build a social media stats tracker display using the Wi-Fi-enabled, Arduino-compatible ESP8266 microcontroller board and seven-segment LED displays. This project is an extension of my YouTube Subscriber Counter, but uses one "brain" to track and display multiple networks. Because these are four-digit displays, you can easily omit the lefthand displays (high digits) for any stat under 10K to bring the cost of components down.

Before attempting this project, you should be generally familiar with uploading new programs to your Arduino board and installing code libraries, both of which you can learn for free in my Arduino Class (instructables.com/class/Arduino-Class), though you really don't have to understand much of the actual Arduino code to get this project running.

1. TEST YOUR ARDUINO

Make sure you've got your Arduino software set up properly to program the board you're using, which in my case involves installing the SiLabs USB driver and installing ESP8266 board support. Find detailed instructions at instructables.com/id/Social-Stats-Tracker-Display-With-ESP8266.

2. PREPARE DISPLAYS

Follow the assembly instructions for your seven-segment displays, and also solder headers onto your ESP8266 board if it didn't come with headers pre-soldered.

For your microcontroller to tell the different displays apart, you'll need to change their addresses. Do this by applying a blob of solder to bridge the labeled pads on the back of the board. The board as-is has an address of **0x70**, and if you bridge the pads labeled A0 (Figure Ⓐ), it has an

address of **0x71**. Here's how the displays in my circuit/code are configured:

0x71 high digits Twitter, A0 shorted
0x70 low digits Twitter, no alteration
0x74 high digits Instagram, A2 shorted (not installed yet)
0x73 low digits Instagram, A0 and A1 shorted
0x75 high digits Instructables, A0 and A2 shorted (not installed yet)
0x72 low digits Instructables, A1 shorted

3. BUILD AND TEST THE CIRCUIT

I like to create a solderless prototype first (Figure Ⓑ), then solder it once it's working. I'm using a NodeMCU board, but any ESP8266 board with the I²C pins exposed will work fine. In my case that's D1 and D2 (Figure Ⓒ). Connections are as follows:
NodeMCU D1 to displays C (clock)
NodeMCU D2 to displays D (data)

TIME REQUIRED:
4–6 Hours

DIFFICULTY:
Intermediate

COST:
$45–$110

MATERIALS
- » **Medium shadow box, 5"×7", 1" deep** such as Amazon #B01GG6U74A amazon.com
- » **NodeMCU ESP-12E microcontroller board** such as Amazon #B01O01G1ES. Or use your favorite ESP8266 board; some require an additional 3V FTDI programmer to upload programs.
- » **Seven-segment LED displays, 4-digit, with I²C backpack (up to 6)** Adafruit #880 adafruit.com
- » **Proto board** Adafruit #591 or similar
- » **Female/male header wires**
- » **Micro USB cable** charge + sync, not charge-only
- » **USB power adapter (optional)**
- » **Solderless breadboard and jumper wires (optional)** for prototyping, Adafruit #239
- » **Template** free download at instructables.com/id/Social-Stats-Tracker-Display-With-ESP8266

TOOLS
- » **Soldering iron and solder**
- » **Printer and paper**
- » **Scissors**
- » **Cutting mat or scrap cardboard**
- » **Illustration board or more scrap cardboard**
- » **Flush diagonal cutters**
- » **Desoldering braid or solder sucker** for mistakes
- » **Utility/craft knife**
- » **Tape**
- » **Metal ruler (optional)**
- » **Computer with Arduino IDE** free download at arduino.cc/downloads

NodeMCU Vin to displays + (power)
NodeMCU GND to displays – (ground)

Download the project code found at the Instructables page (big thanks to Brian Lough for writing all those awesome Arduino libraries) and follow the instructions there for inserting your own Twitter API bearer token, usernames for Twitter, Instagram, etc., and Wi-Fi network name/password.

Now upload the code to the Arduino and peep your tweeps!

4. ASSEMBLE IN SHADOWBOX
Download the paper template at the Instructables page or create your own 5"×7" artwork to fit in the front of the shadowbox frame (Figure **D**). This template also contains an alignment helper, which you can tape the displays directly to, or use as a reference for taping the displays to the back of your front-facing art. The thickness of your paper will determine how much light shines through (and circuit sturdiness), so feel free to play around with the combo of layers and paper thickness that works best for you.

I soldered my two-display counter (Twitter) to a proto board to keep the numerals aligned, and plugged female header wires into the single-display headers. The NodeMCU board gets its own small proto board, where I soldered more wires to plug into all the various displays (Figure **E**). This keeps the project modular in case I want to take it apart or add more displays later.

5. CLOSE IT UP
Use cardboard or other stiff material to brace against the back of each display (Figure **F**) before installing the back cover. Use scissors to cut a notch for the USB cable (Figure **G**). It took me a few tries to get the displays pressed evenly against the artwork/glass to create crisp digits. If they're not pressed tight, the numbers will appear blurry.

MAKE 'EM COUNT
Display your project and track your growth! I love having these displays hanging above my workbench to inspire me to keep making and sharing fun projects. I'd be delighted to see your trackers and to hear your thoughts and questions. Find me @bekathwia on YouTube/Twitter/Instagram. ✇

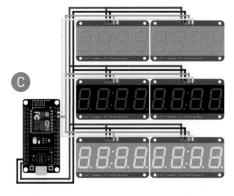

STAR WARS®
THE FORCE UNLEASHED

RAHM KOTA
—LIGHTSABER—

Written and photographed
by Darrell Maloney

Not Your Father's
Lightsaber

Jedi or Sith? Design
and build your own
fully operational
3D-printed saber and
choose your path

DARRELL MALONEY,
better known on YouTube
as The Broken Nerd
(makezine.com/go/broken-
nerd), is a maker and
hobbyist who is passionate
about 3D modeling, 3D
printing, and prop making.

A

B

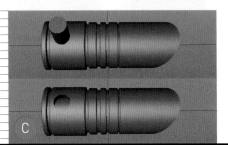

C

I've always wanted to build my very own *Star Wars* lightsaber, but with no access or knowledge of how to operate a metal lathe, I opted to 3D model and 3D print one instead.

Unlike some other printed sabers, I wanted mine to be fully functional — lights, sounds, switches, and sensors. This required me to look into how the lightsaber prop builders' community assembles their sabers and apply that methodology to my 3D-printed version. Here's how I made my Rahm Kota saber using Cinema 4D; you can use similar tools to do this in any CAD program.

1. 3D MODELING

You'll start by creating a Cylinder object for all three parts of the hilt: the mid hilt, top hilt, and bottom hilt.

1a. Mid hilt

The mid hilt needs to have an outer diameter of 38mm and an inner diameter of 28.5mm; this provides enough room for the lightsaber electronics kit. (We will keep this inner diameter at 28.5mm throughout the hilt, except at the very top and the very bottom.)

Use the Extrude tool in your CAD program to shrink the top of the mid hilt

inward, to an outer diameter of 35mm, and then upward about 5mm high (Figure A) — this will provide a lip and flange to connect to the top hilt piece.

At the bottom of the mid hilt, extrude inward to 37mm and extrude downward 11.6mm — this lip will sit inside the bottom hilt.

Finally, use the Cut or Boolean tool to cut a 4.7mm-diameter hole into the bottom lip you just extruded (Figure B). This hole will accept a 5mm Allen (hex) setscrew to hold the mid hilt and bottom hilt together.

1b. Top hilt

The top hilt has an outer diameter of 40.6mm, and a few minor inward and outward extrudes to create details.

For the upper 45mm of the top hilt, shrink the inner diameter to 26mm, so that a standard 1"-diameter lightsaber blade will fit snugly in the top of the hilt.

Using the Cut or Boolean tool, cut a 12.5mm hole about 13.5mm from the bottom of the top hilt (Figure C), for the on/off switch to sit in.

Finally, cut a 4.7mm hole on the backside of the top hilt, about 20.8mm from the top

(Figure D). This hole will accept a second setscrew to securely hold the lightsaber blade in place.

Style your saber however you want — I cut off the top at an angle, then cut an arc out of the tip to match Jedi Master Rahm Kota's weapon.

1c. Bottom hilt

The bottom hilt is 38.6mm long, with an outer diameter of 45mm. The inner diameter is 38mm — larger than the rest of the saber, because the speaker at the bottom of the lightsaber kit will sit in this space.

Style the bottom end as you wish; I used large and small dome shapes to form a sort of pommel. Use the Cut or Boolean tool to create holes where you wish, so that you can hear sound from the speaker (Figure E).

Finally, cut another 4.7mm hole, matching the one in the mid hilt, for the setscrew that fastens the two together (Figure F).

2. PRINTING

Export the models as STL files and open them in a slicer of your choice (Cura, Simplify3D, Meshmixer, etc.).

For strength you can print the lightsaber at 100% infill; I recommend no less than 50% infill (to support the weight of the lightsaber blade and kit) and 0.2mm layer height.

3. PREP AND PAINT

Sand all parts, starting with 100 grit sandpaper and ending with 220 grit, or until desired smooth surface is achieved. Sanding the mid hilt is optional as the leather hide will cover this area.

Before applying primer/filler, use the Dremel to create scratch marks to simulate weathering or battle damage.

Once the primer/filler is dry, use the airbrush to apply Alclad black gloss; once that's dry, apply Alclad aluminum (Figure G). Let all the parts thoroughly dry, then apply Maskol to areas where a chipping effect is desired, and use Frog Tape to mask off larger areas that need to remain silver (aluminum). Now apply Model Masters Gun Metal to all exposed areas. Once dry, peel off the Frog Tape and the Maskol to reveal aluminum areas.

Mix brown and black acrylic paint, apply to the hilt, and wipe off excess paint, leaving the remaining paint in the detail areas. This will create a weathered look (Figure H).

4. FINAL ASSEMBLY

Place the lightsaber electronics kit in the hilt and use 5-minute epoxy to glue the top hilt to the mid hilt. Use a setscrew to attach the bottom hilt to the mid hilt.

Cut a long strip of the leather hide, and using contact cement, wrap the mid hilt (the grip). Now dry-brush black acrylic along the edges of the hide to create a weathered effect.

TIME REQUIRED:
Modeling: 2–5 Hours
Finishing: A Weekend

DIFFICULTY:
Advanced

COST:
$150–$180

MATERIALS
- » **Printer filament, PLA, 2.2lb roll**
- » **Testors Model Master lacquer paint, Gun Metal, ½oz bottles (2)**
- » **Alclad II lacquer paints, 1oz bottles: Gloss Black (1) and Aluminium (1)**
- » **Humbrol Maskol, 28ml bottle**
- » **Sandpapers: 100, 150, and 220 grit**
- » **Rustoleum Filler Primer, spray can**
- » **Outer Rim Sabers Lightsaber Electronics Kit** includes soundboard, 3W LED, rechargeable 800mAh lithium battery, power switch, 12mm momentary switch, and 23mm 8Ω speaker, etsy.com/shop/OuterRimSabers
- » **Lightsaber blade, 1" outer diameter** Find your favorite on Etsy, or make your own.
- » **Leather hide, brown, 12"×24"**
- » **Acrylic paints, 2oz: black (1) and brown (1)**
- » **Contact cement**
- » **Frog Tape**
- » **5-minute epoxy**
- » **Hex head setscrews, 5mm diameter (2)**

TOOLS
- » **3D printer**
- » **High-speed rotary tool** e.g., Dremel
- » **Airbrush (optional)** You can substitute spray paints for the bottled lacquers.
- » **X-Acto knife**
- » **Self-healing cutting board**
- » **Hot glue gun**

[+] Watch the video at youtube.com/watch?v=bCc5Lx_xcmc and see more Broken Nerd at youtube.com/user/makemagazine/playlists.

USE THE FORCE

Place the blade into the hilt and fasten it with the other setscrew. Now that you're ready to ignite your first lightsaber, push the On button and enjoy! Upon slight impact your saber will make clashing sounds.

I also designed a cool display stand for my saber, and I've been fooling around with a kyber crystal chamber for a display-only model.

Now you have the tool to become a Jedi Master or Sith Lord. May the Force be with you! ●

TIME REQUIRED:
A Weekend

DIFFICULTY:
Easy

COST:
$40–$60

MATERIALS
- » Wood dowel, 1" diameter, 18" long
- » Deck screws, 3½" (1 box)
- » Pine board, 1×10, 14" long for the base
- » Pine boards, 2×4, 5" long (4) and 11" long (1) for risers and crosspiece
- » Wood screws, #8×1½" (8)
- » Iron pipe flange, 1¼" diameter
- » Machine screws, #10-24×1" (3)
- » Wood screws, flat head, #10×¾" (4)
- » Aluminum shaft collar, 1" McMaster-Carr #9946K24, mcmaster.com
- » Ball bearing, open, 1" ID, 2" OD R16 type, McMaster #60355K509
- » Aluminum bars, 1/16"×½": 9½" long (1), 10" long (1), and 7¾" long (2)
- » Wooden balls, 2" diameter (2)
- » Steel hex head bolts, ¼-20 full thread: 6" (2) and 3" (1)
- » Plastic washers, low friction, ¼" (8) McMaster #2796T11
- » DC motor, 6V–20V, about 1 ¼" dia. by 2 ½" long, with 3mm shaft Search online for an "electric drill motor."
- » Aluminum mounting hub for 3mm shaft, #4-40 holes Pololu #1078, pololu.com
- » Machine screws, #6-32×¾" (2) for shaft collar string anchors
- » Brads, 18 gauge, ¾" long (4) hub to shaft attachment
- » Nuts, ¼-20 (12) for linkage
- » Wood screws, #8×¾" (2) brush to stand attachment
- » Mason twine, #18, 2"
- » Alligator test clips (4)
- » Batteries, 9V (2)
- » Thread-locker adhesive
- » Gorilla Glue or epoxy

TOOLS
- » Electric drill with standard bits
- » Large drill bits: 1¹⁄₁₆" twist bit and 2" Forstner bit
- » Tap and die set with hole taps: ¼-20, #10-24, and #6-32
- » Vise
- » Pliers, heavy duty (2)
- » SAE socket set
- » Tack hammer
- » Medium sandpaper
- » Safety glasses

James Watt

and the

Flyball Governor

Build the ingenious mechanical regulator that made steam engines run on time

Written by William Gurstelle

In 1764, scotsman James Watt grew thoughtful as he tinkered with the machinery associated with his job as a steam engine technician. The main problem with those early steam engines, called Newcomen engines, was that they were terribly inefficient (Figure ). They were a step up from draft horses, but Newcomen engines went through coal like a starving man on a Christmas ham.

This inefficiency, thought Watt, was simply unacceptable, and he believed he could do better. When Watt put his mind to a task, he was a nearly unstoppable force. Besides his preternatural drive and determination, he possessed, as English historian Samuel Smiles said, "a keen eye for details, with which he combined a comprehensive grasp of intellect."

In fairly short order, Watt made radical improvements to the early steam engines, boosting their efficiency and utility. Those improved Watt engines changed the world, leading to the Industrial Revolution in England, America, and beyond.

But Watt wasn't finished yet. One of the many engine-related inventions for which Watt is credited is the *flyball* or *centrifugal governor*. In the early days of the Industrial Revolution, most engine operators in mills and mines needed to keep their machines turning at a constant rate of speed, irrespective of the load placed on them. Generally the speed of an engine increases when the load decreases, so, for example, if a pump removing water from a mine shaft needed to operate at 70rpm, then the operator had to manually open or close the steam valve whenever the amount of water in the mine shaft changed.

In 1788, Watt began to think about a way to make this happen automatically. His solution was the flyball governor (Figure 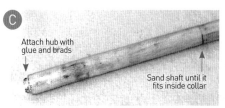).

The flyball governor is based on the idea of a *feedback control loop*. It works like this: Weighted balls are linked via hinged arms to the shaft of a steam engine. As the engine turns faster, the hinged flyballs fly apart. But, as the balls separate, a linkage causes the throttle on the steam engine to close.

Less steam means the engine slows down and the flyballs come back together. But, when they get too close to one another, the linkage causes the throttle to reopen and the engine speeds up again, and thus an endless loop of engine speed adjustment ensues. If designed well, the flyball governor maintains a fairly constant engine speed, no matter what the load.

While Watt didn't originate the concept of the centrifugal governor (Christiaan Huygens invented one for clocks, see "Poetry in Motion," page 40) or the feedback loop, his is one of the first important examples of the idea in use in large machines.

In this edition of Remaking History, we make a James Watt-style flyball governor. Since few of us have steam valves and boilers readily available for a project, we'll use the electricity from a couple 9-volt batteries in place of the boiler, and a pair of aluminum brushes instead of a steam valve. This simplified, on-off rendition of Watt's world-changing wonder is cheap to build, fairly easy to make, and great fun to show off to friends and family.

BUILD AN ELECTRICAL FLYBALL GOVERNOR

There are four main components of this flyball governor project: the shaft, the base, the flyball linkage, and the control brushes.

The Shaft
1. Use the sandpaper to reduce the diameter of the dowel until the aluminum shaft collar slides easily along its full length. Then, drill a $5/16$" hole through the dowel $2\frac{1}{2}$" from the top.
2. Place a small amount of glue on the bottom end of the dowel. Place the aluminum mounting hub on the dowel and center it as perfectly as you can (Figure ). Let the glue dry. Once dry, hammer the brads into the holes on the mounting hub to permanently secure it to the dowel.

The Base
3. Drill three $5/32$" holes, 120° apart, in the wall of the pipe flange. Cut bolt threads into each hole using a #10-24 tap. Screw the #10 machine screws into the holes (Figure 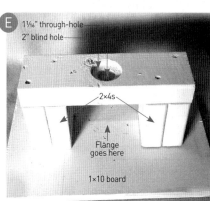). You can find instructions for tapping threads in metal at makezine.com/2011/03/22/skill-set-the-basics-of-tap-and-die.
4. Using the Forstner bit, cut a 2"diameter flat-bottomed "blind" hole, $\frac{3}{8}$" deep, in the center of the top surface of the wooden support as shown in Figure 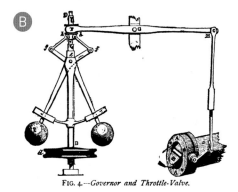. Then, using the center mark made by the bit, cut a $1\frac{1}{16}$" hole through the center of the blind hole.
5. Follow the instructions in Figure E to build a wooden shaft support using the 2×4 boards and the 1×10 board.

FIG. 4.—*Governor and Throttle-Valve.*

Attach hub with glue and brads

Sand shaft until it fits inside collar

Drill and tap three #10-32 holes in pipe flange

$1\frac{1}{16}$" through-hole
2" blind hole

2×4s

Flange goes here

1×10 board

WILLIAM GURSTELLE's new book series *Remaking History*, based on this magazine column, is available in the Maker Shed, makershed.com.

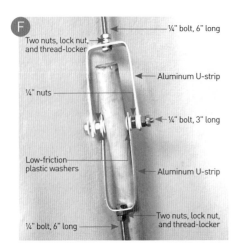

Two nuts, lock nut, and thread-locker
¼" bolt, 6" long
Aluminum U-strip
¼" nuts
¼" bolt, 3" long
Low-friction plastic washers
Aluminum U-strip
¼" bolt, 6" long
Two nuts, lock nut, and thread-locker

F

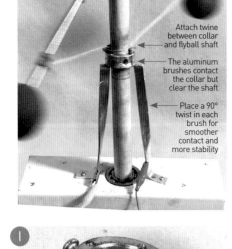

G

H

Attach twine between collar and flyball shaft

The aluminum brushes contact the collar but clear the shaft

Place a 90° twist in each brush for smoother contact and more stability

I

[+] Watch video of the flyball governor in action, and share your build, at makezine. com/projects/remaking-history-flyball-governor.

6. Place the modified pipe flange in the middle of the 1×10 board, centered directly under the 1¹⁄₁₆" hole. Fasten the flange with the #10×¾" wood screws.

The Flyballs and Linkage

7. Drill ⁷⁄₆₄" holes in the center of the 9½" and 10" aluminum strips.
8. Bend the 9½" and 10" aluminum strips into U shapes as shown in Figure **F**.
9. Using a ¹³⁄₆₄" bit, drill and then tap a ¼–20 threaded hole, ¾" deep, in each wooden ball.

Control Brushes and Shaft Collar

10. Use the vise and pliers to bend the 7¾" aluminum strips into the shape depicted in Figure **G**. Then, add a 90° twist as shown in Figure **H**. These are the brushes.
11. Drill two ⁷⁄₆₄" holes 180° apart on opposite sides of the shaft collar for twine attachments. Cut 6-32 threads in the holes. Screw the 6-32×¾" bolts into the threaded holes (Figure **I**).

Putting It All Together

12. Insert the motor into the flange opening.
13. Place the bearing in the 2" blind hole.
14. Insert the shaft, hub down, through the center of the bearing. Connect the hub to the motor shaft using the setscrew and Allen wrench that came with it. Check and adjust to make sure the shaft spins true and easily. Then fasten the motor in the flange by tightening the #10 bolts.
15. Slide the aluminum shaft collar onto the wooden shaft.
16. Attach the aluminum motor brushes to the wooden support frame as shown in Figure H using the #8×¾" wood screws. Gently bend the aluminum as needed so that it just clears the shaft, but supports the shaft collar as shown in Figure H.
17. Assemble the flyball linkage (the 3" hex bolt, nuts, and plastic washers) as depicted in Figure F and attach to the wooden shaft as shown in Figure H.
18. Connect a 10" length of mason twine between the attachments on the shaft collar and the outboard shaft nuts as shown in Figure H.
19. Use the alligator clips to complete an electrical circuit between the motor, batteries, and brushes as shown in Figure **J**. (If the motor contacts are under the flange, splice in a wire to the contacts so you can connect the alligator clip.)

HOW YOUR FLYBALL GOVERNOR WORKS

When you complete the electrical connection, the motor shaft will rotate. As it rotates, centrifugal force causes the balls to fly outward, causing the mason twine to lift the aluminum shaft collar off the brushes, disconnecting the circuit and causing the motor to slow down. But, the slower rotation means the flyballs will move inward again and the shaft collar will lower once again upon the aluminum brushes, remaking the electrical connection. When this happens, the motor speeds back up.

This is an example of a feedback control loop: The unending cycle of making and breaking the electrical contacts ensures the motor rotation speed stays within a narrow band, providing more or less constant RPM, irrespective of the load placed upon it. James Watt would be proud!

OPERATION NOTES

» Wear safety glasses whenever the shaft is in motion.
» For safety, do not let the shaft turn at speeds greater than 90rpm. You can adjust the rotational speed by making the mason twine lengths or aluminum brushes longer or shorter.
» Secure all threaded connections with thread-locker adhesive to prevent nuts and balls coming loose.
» The make-and-break style circuit results in tiny sparks as the machine rotates. These sparks will eventually leave a nonconductive residue on the aluminum brushes and collar, which must be sanded off periodically for best performance. ●

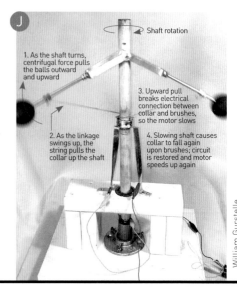

J

Shaft rotation

1. As the shaft turns, centrifugal force pulls the balls outward and upward

3. Upward pull breaks electrical connection between collar and brushes, so the motor slows

2. As the linkage swings up, the string pulls the collar up the shaft

4. Slowing shaft causes collar to fall again upon brushes; circuit is restored and motor speeds up again

William Gurstelle

1+2+3 Simplest Soldering Station

Written, illustrated, and photographed by Chris Connors

When students do soldering in my classroom, they need a tool to hold their work and protect the tables from getting burned. I've tried adjustable soldering stands, but they're pricey, they've got no protective base, and they're too easy to take apart, so the parts get lost.

So I made my own. This simple design has served well for years and is made largely from scrap. It uses an Altoids tin for storage of solder, desoldering braid, and extra iron tips. The base protects the table, and it fits in a plastic tub with safety glasses, irons, and a solder sucker.

1. BUILD THE BASE

Lay the 2×4 across the plywood base, 2⅝" from the back edge (to leave room for the mint tin). Attach the boards from the bottom with two 1½" screws.

2. ADD BINDER CLIPS

Attach 1 or 2 binder clips to the 2×4 using ½" screws and washers. I like to put one close to the edge, and one set in a bit. Add a third clip on the plywood base to give a variety of locations to secure the project you're soldering (or desoldering).

3. ADD THE MINT TIN

Affix the tin on the back section with its hinge facing the 2×4, using two ½" screws and washers. When opened, the lid rests against the 2×4, so it's less likely to break off.

USE IT

Open the binder clip and secure your work. Soldering a DC motor? Put the motor shaft into the clip on the edge. A circuit board? Put it in the clip of your choice. And for larger boards, the support of the lower area is useful. ✐

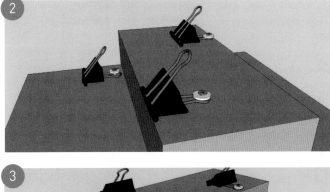

TIME REQUIRED:
30–60 Minutes

DIFFICULTY:
Easy

COST:
$0–$5

YOU WILL NEED
» Plywood, ¾"×8"×10"
» 2×4 board, 8" length
» Drywall screws, ½" (5) and 1½" (2)
» Washers (5)
» Mint tin
» Binder clips, medium size
» Saw
» Screwdriver

CHRIS CONNORS
teaches programming and new technologies in the art department of Martha's Vineyard Regional High School in Massachusetts. He makes things and cheers others on to greater heights.
chrisconnors.com

BOB KNETZGER

is a designer/inventor/ musician whose award- winning toys have been featured on *The Tonight Show*, *Nightline*, and *Good Morning America*. He is the author of *Make: Fun!*, available at makershed. com and fine bookstores everywhere.

Show 'N' Glow
Toy Display

Upcycle a vintage plaything as a showpiece, comix rack, and reading lamp!

Written, photographed, and illustrated by Bob Knetzger

TIME REQUIRED:
3–4 Hours

DIFFICULTY:
Easy

COST:
$50 + Toy

MATERIALS
» **Vintage toy of your choice**
» **Headlamp** an inexpensive one from the hardware store is fine
» **Acrylic sheet, white, ¼"**
» **Acrylic sheet, clear, ⅛"**
» **Nuts and bolts (2)**
» **Small plastic tubing** for spacers to fit around bolts

TOOLS
» **Soft pencil**
» **Saw, fine toothed** for cutting plastic. You can have the plastics shop cut your acrylic to size. For the cutouts, and the PVC, I used a band saw but you could use a jig or scroll saw with a fine-toothed blade.
» **Drill** I recommend using special plastic-drilling bits.
» **Hand files, small**
» **Strip heater**
» **Screwdriver**
» **Laser cutter (optional)** perfect for cutting the acrylic

[+] See more photos and share your build at makezine.com/projects/ toy-inventors-notebook- show-n-glow

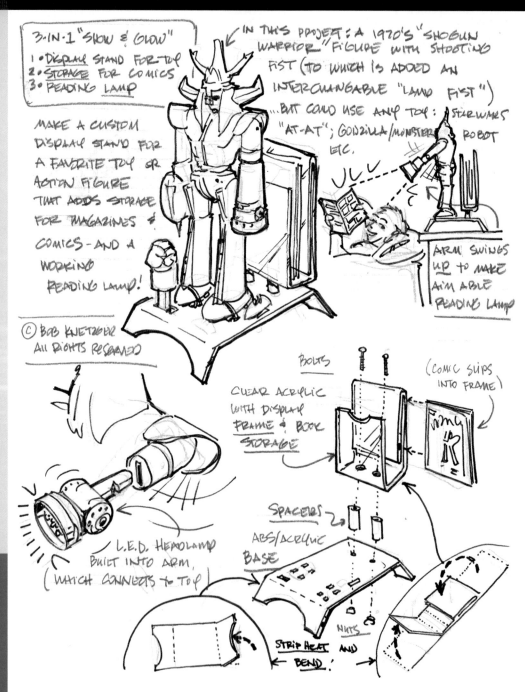

It's fun to see favorite old toys displayed or used in new ways. Old game boards can be upcycled into spiral bound notebook covers. Toy catalogs are laminated to make wallets. I wanted to come up with a new way to display a vintage Japanese robot toy and put him to work, too. Here's a three-in-one design for a "Show 'N' Glow" toy display stand with a reading lamp combined with comic book storage in the back (Figures  and )!

TRANSFORMING A TRANSFORMER

The *Shogun Warriors* were a popular line of Japanese robot toys imported by Mattel in the mid-1970s. Based on a kid's cartoon show in Japan, they featured characters like Raydeen, who transformed into a birdlike spaceship. The imported toys' wild designs seemed very exotic, with Japanese labels and markings. Standing 2 feet tall, the *Shogun Warriors* bristled with launching rockets and spring-loaded punching fists. (Visit the project page at makezine.com/projects/toy-inventors-notebook-show-n-glow to see a *Shogun Warriors* toy commercial and the Japanese show intro.)

Today, these 40-year-old toys command collector prices on eBay. I wanted to make sure I didn't compromise the toy's value by modifying it, but I did do one upgrade. *Shogun Warrior* fans can buy new labels to replace the long-ago-faded originals. With fresh, unfaded labels Raydeen is now ready for his close up!

> **TIP:** Repainting or modifying a very valuable old toy can be a costly mistake. Check first! And before you use a solvent or cleaner, test it first in an inconspicuous place on your vintage treasure.

LIGHT WEAPONRY

I also hacked a camping headlamp and made a plastic collar for it that fit snugly onto Raydeen's adjustable fist-shooting arm. That way Raydeen becomes an aimable reading lamp without any permanent modifications (Figures  and).

The display stand is a simple two-part design. The base is made from thermoformed ¼" white acrylic with holes drilled to provide storage for the robot parts. The comic book rack is clear ⅛" acrylic, thermoformed into a U-shape. The two parts are joined together with nuts and bolts

with PVC tubing sleeve spacers (Figure).

I added a laminated full-size copy of the cover of *Shogun Warrior* comic #1 to the back of the comic storage (Figure ).

This particular headlamp makes a great reading light with three brightness settings — and flashing red danger lights! Raise Raydeen's arm to aim the reading lamp.

MAKE IT YOUR OWN

This concept could work with almost any favorite action figure or toy, new or old. Adapt this basic design to create your own Show 'N' Glow. An old *Star Wars* AT-AT could find new life as a reading lamp with some bright LEDs mounted on the chin guns. A roof-mounted searchlight on a *Teenage Mutant Ninja Turtles* van would make a swell night light. You get the idea ...

Your project will dictate the actual dimensions and details, but here are some construction tips:

» While you're at the plastic store, get some special drill bits made for drilling in plastic. They have a steeper 60° point and specially ground flutes that won't grab or crack the plastic.

» Make a quick full-size sketch layout of the mounting details for your toy. I wanted square holes to accept the small wheels on the bottom of the robot's feet, as well as storage for the extra bird missiles and punching fist. I used a piece of ¼" foamcore board (Figure ) to check the sizes and locations of the holes and the amount of bend for the back and front "legs" of the stand (Figure 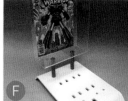).

» If your acrylic sheet has a protective film, mark the locations and sizes of holes in soft pencil on the film (Figure ), then cut and drill away. Later, be sure to remove the film before heat bending.

» To make square holes, first drill out the center, then finish with small hand files (Figure 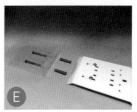). (If you have access to a laser cutter, this project is a natural!)

» Use a strip heater to make your bends. Mark the bottom side of the acrylic in pencil to keep track of which side is which. The heated side is always the *outside* radius of the bend, where the plastic has to stretch and bend the most (Figure ). For more on making and using a strip heater see makezine.com/projects/toy-inventors-notebook-bend-a-battery-box.

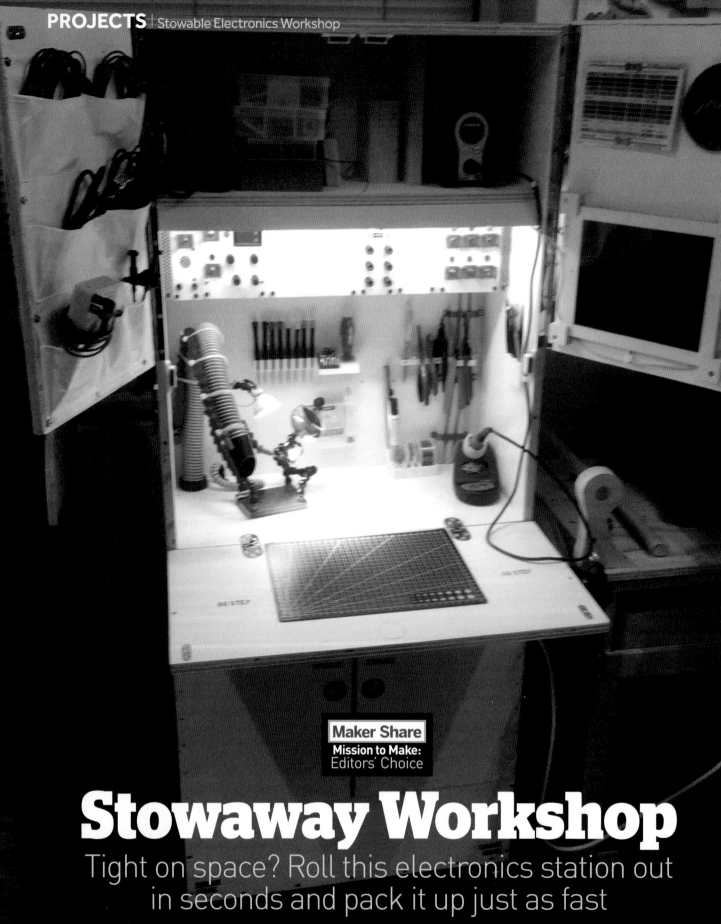

Maker Share
Mission to Make:
Editors' Choice

Stowaway Workshop

Tight on space? Roll this electronics station out in seconds and pack it up just as fast

Written and photographed by Morten Nisker Toppenberg

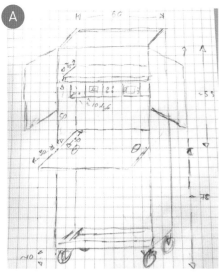

A

B

TIME REQUIRED:
75–100 Hours

DIFFICULTY:
Intermediate

COST:
$150 + Tools

MORTEN NISKER TOPPENBERG is a maker, tinkerer, fixer of broken toys, amateur photographer, and *Doctor Who* fan. He lives in Egå, Denmark with his wife and their two boys. Find him at nisker.net

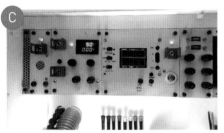

C

D KEEP CLEAR / NO STEP

E WARNING FUME EXHAUST

F

G

H

I

I HAVE A SMALL WORKSHOP WHERE I BUILD ALL KIND OF THINGS, BUT SOLDERING AT A WOODWORKING BENCH is not optimal. I wanted an electronics workbench — but with space at a premium, I had to get creative with storage without sacrificing ease of use.

The overall design (Figure A) is a cross between an airplane trolley and a tool cart. By putting it on wheels I could store it sideways, saving space, and also move it wherever I needed in the shop (Figure B). It was also very important that it be fast to set up and repack — ready to go as soon as it was opened and powered up.

I'm a big fan of Tom Sachs' art, in particular his short movie *Space Program*, so the design aesthetic with the painted plywood is very much borrowed from him. This gives the whole thing a sort of used and worn look that you would perhaps see on a maintenance cart on a station in the outer rim of the galaxy.

It's a quite capable little workshop for electronics, with a soldering station and fume extractor (with carbon filter), a constant-current constant-voltage variable power supply, and a simple oscilloscope (Figure C). The tools I use most are in plain view. Assorted parts like heat-shrink have their place on the upper shelf, and there's a storage area below the power distribution that can pack quite a bit of stuff.

The main carcass is made from 15mm plywood that's just screwed together. To get the Tom Sachs look, it is painted before it is cut. The paint is a non-glossy white so it picked up a lot of scuff marks from being cut and assembled, as intended. The stenciled letters are made on a vinyl cutter and transferred using transfer sheet. The font is Boston Traffic (Figures D and E).

I used a lot of found materials for this project, like the vacuum cleaner hose for the fume extractor and the painted, grimy aluminum used to cover the electronics. I've also been rummaging through the electronics graveyard and parts bins at my local makerspace — a good source of parts with an "old" look.

AHA! MOMENT
Soon after mounting the shelf and fold-down table, I rigged up the lighting and fume extractor and started soldering — using the project to build itself!

UH-OH! MOMENT
After I mounted the wheels, the main carcass was tipping over extremely easily. I eventually put 25kg (55lbs) of crushed rock in the bottom, which lowered the center of gravity and made it very stable.

And then there are the 3D-printed parts. I designed and printed a lot of different bits and bobs to get the look and functionality I wanted: tool holders (Figures F and G), flush-mounted cupboard pulls, NASA-inspired switch guards (Figure H). The most complex was probably the fume extractor (Figure I). You can download them all for free at thingiverse.com/nisker/designs.

[+] Follow and share at makershare.com/projects/stowable-electronics-workshop

Fantastic Plastic

Create a clever cosplay crown using Cameo paper cutouts and an easy plasticizing technique

Written and photographed by Julia Jameson Barton

TIME REQUIRED:
1–2 Hours

DIFFICULTY:
Easy

COST:
$20–$40

MATERIALS

» **E-Z Water plastic pellets** Woodland Scenics #C1206, Amazon #B000BL8JNI
» **Heavy cardstock**
» **Glue stick**
» **Paints** in the colors you desire. Acrylic paints and spray paints can be used. Nail polish looks great for details. Here I used copper and black spray paints, and gold powder.

TOOLS

» **Desktop electronic cutter (optional)** I use a Silhouette Cameo. You can also cut the paper by hand for this project, but the electronic cutter makes it easy to create very fine and accurate details.
» **Non-stick aluminum foil**
» **Cookie sheet or baking pan** with a rim all the way around; large enough to fit your cutout in
» **Oven** I use a dedicated toaster oven for crafts. If you use your kitchen oven, clean it afterward, before preparing food.
» **Heat resistant gloves**
» **Tweezers**
» **Scissors**
» **Marker**
» **Heat embossing tool or heat gun** to tidy up the edges after dipping paper. A hair dryer isn't hot enough.
» **Hot glue gun (optional)**

JULIA JAMESON BARTON (The Juliart), a lifelong artist, is an inventor and product developer. She works in countless mediums, creates commissioned art, and loves helping others to find their medium.

FOR PROPS AND COSPLAY THERE ARE MANY THERMOPLASTIC MATERIALS ON THE MARKET — but some are not budget friendly. I've been a maker all my life and a product developer for 20 years, and most of my creations are made from materials and processes that were intended for something entirely different. This is one of them.

E-Z Water is a thermoplastic material used for modeling miniature water scenery, but uniquely enough, it's meant to be poured onto substrates and then harden, so its properties lend themselves to this project and many others.

Let's create a cool crown for your next cosplay — by cutting paper patterns and then saturating them in the thermoplastic to make faux metal filigree. I use my Silhouette Cameo, but you can use any cutter you like.

1. CREATE YOUR PATTERN
Use a drawing program, or just freehand it. In Silhouette Studio, choose flourish designs or other images, then resize and Weld them together to create a crown design you like (Figure A). You can erase any undesirable areas created by the overlapping images.

Don't worry about designing a whole crown at once. You can create it in sections and glue them together later with more E-Z Water or hot glue.

2. CUT THE PATTERN
Cut the pattern on your electronic cutter (Figure B). Check the fit and if necessary, resize and recut.

> **TIP:** For strength, cut 2 identical pieces and glue them together before dipping.

3. MELT THE PLASTIC
Preheat your toaster oven to recommended temperature on the E-Z Water package.

Line the cookie tray with non-stick foil and pour in the E-Z Water beads. Place tray in oven and let the beads melt (Figure C), paying close attention (no kids, no pets, you get the picture).

> **TIP:** The beads are a light amber color. To create colored material, you can mix in powder colors, mica powders, etc. while the plastic is in this liquid state. The material can be remelted again and again.

4. DIP YOUR PATTERN
With your gloved hand and tweezers, submerge your paper cutout in the melted tray of E-Z Water (Figure D).

In a clean and protected work area, lay out another large piece of non-stick foil and a potholder to set your hot pan on. Gently remove the pan, then use your tweezers to lift the paper, being careful to drip only in the pan and not on your hands. Quickly lay the cutout down on the foil.

5. TRIM AND SMOOTH
Your dipped piece will cool fairly quickly depending on room temperature. Now you can trim away excess material and drips.

Here's where the heat tool comes in (Figure E). Hold your piece above the foil about an inch, then carefully apply heat to soften, melt, and smooth the cut edges or blemishes. Do this till you get your desired look on front and back. I prefer to heat the back first, then move to the front.

6. FORM THE SHAPE
Now warm and form your crown (Figure F). Use heat-resistant forms, and wrap them with non-stick foil if needed. Heat the material a little at a time and gently push into place, then hold till cool.

7. PAINT
Follow the directions for your paint type. I paint my pieces with black first, then begin layering color (Figure G). Splatter paint for cool effect. Wipe it off or texture it using crinkled-up paper or foil. Try textured spray paints, or add phosphorescent glow powders for a truly unique wearable art.

CROWNING ACHIEVEMENT
You can embellish your crown with gems, metal wire, and other details by gluing them on. Create your own straps or headbands, or reconfigure store-bought ones to finish out your thermoplastic paper creation.

The E-Z Water material is versatile — because it melts to a liquid it can saturate paper and natural fabrics. (By comparison, Worbla is pretty costly and basically only molds to itself.)

It's also fairly durable. It will crack if you're rough with it, but I still have pieces I created over 6 years ago! Have fun with it. ●

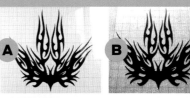

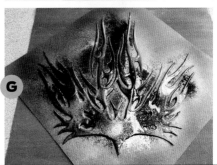

DJI MAVIC AIR $799
and PARROT ANAFI $600
dji.com and parrot.com

These two compact drones bring a new class of portability to aerial videography. Similarities: Both feature folding arms, offer roughly comparable video quality, and use improved Wi-Fi for control with matching ranges — we got about a mile away in tests.

WHY WE LIKE THE MAVIC AIR:
Two main reasons: collision sensors (both front and rear), and speed. DJI has made headway with obstacle avoidance, and for anyone who's flown into a tree or the side of a building, they'll be thankful for this. Meanwhile "Speed Mode" gives the pilot a fast-as-heck 43mph top speed (be warned, however — the collision sensors don't work with this setting).

The Mavic Air also generally feels more rugged than the Anafi, and offers more flight functions than the Parrot, all for free; Anafi's visual tracking/follow-me option is an in-app purchase.

WHY WE LIKE THE ANAFI:
Factoring in all elements — charging, flying, filming, and even just toting it — the Anafi feels easier to use than the Mavic Air. Its folding configuration and direct-USB charging make it more compact than any other quality drone; all you need for outings is the slim case and remote — there's no bulky external charger anymore. We threw ours into our bag for a few trips "just in case." It's also noticeably quieter, another ease-of-use perk.

The lossless digital zoom gives new perspective options, as does the 90°-upward-rotating lens. We found its battery to last a bit longer than the Mavic Air. Lastly, it costs about $200 cheaper.

WHICH TO GET?
If you want to ensure the safety of your drone and don't mind toting an extra bag with necessary hardware, the Mavic Air is your best bet. But for getting spontaneous shots — often the best type — the Anafi's incredible portability makes it a compelling option. —*Mike Senese*

LITTLEBITS AVENGERS HERO INVENTOR KIT

$150 shop.littlebits.com

Coding sites and electronics kits can often feel *too* educational, the end result uninteresting or intangible. That's where littleBits has always differentiated itself, with playful offerings like the Avengers Hero Inventor Kit, which lets kids (and kids at heart) experiment with hardware and software while customizing an interactive, wearable gauntlet.

The magnetic connectors, which include an accelerometer, LED matrix, and sound effects, and Marvel-themed companion app, are very intuitive. My son, 10, and daughter, 7, needed minimal help as they tackled the character-specific, step-by-step "hero training" — such as stealth mode for Black Widow and lights and sound for Black Panther — high-fiving the plastic Iron Man glove (which I wish included storage for spare Bits) after each success. And the kit definitely got them both to think outside the box — pulling out Nerf armor, walkie-talkies, and assorted accoutrement to incorporate into their play.

What's not to like about a toy that teaches while also igniting imaginations and cultivating creativity? It will be fun to see all that it inspires. —*Laurie Barton*

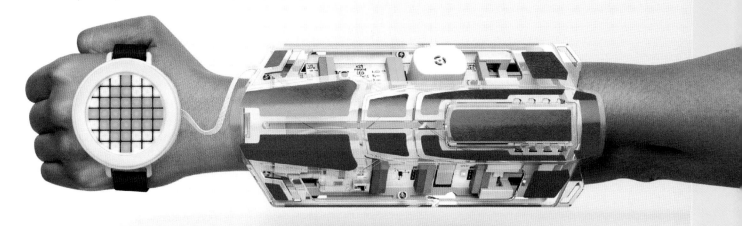

LED NEON FLEX STRIP

$32 amazon.com/dp/B01JZ75QHK

People have been trying to create a new version of neon lights for years. We've seen attempts done by pulling electroluminescent wire through acrylic tubes, CNC milling false plastic structures, and of course, diffusing LEDs.

Usually the LED method comes out on top due to ease of use and lack of overheating, but the method of diffusing is always difficult and often leads to obnoxious bright spots when seen from a distance.

This new product, which you can find on Adafruit, SparkFun, and Amazon, is simply called LED Neon Flex Strip, and it is wonderful. The LEDs are fully diffused, typically the construction is fairly water resistant, it is very flexible, and often it comes with its own control box which allows for simple fading effects and brightness control. It's available in a variety of colors, from warm white to bright pink.

I know I'll be using this stuff for a few projects, it is just so darn pretty! —*Caleb Kraft*

TOOLBOX

DEWALT 20V FRAMING NAILER

$399 dewalt.com

If you're affixing lots of 2×4s to each other (say, for instance, building a house or deck, or even just making a room addition), a framing nailer could be your best friend. Powerful enough to blast structural nails over 3" long deep into wood, these have long been exclusively pneumatic, requiring compressors and hoses to do the job.

The DCN21PLM1 Framing Nailer is DeWalt's latest model, helping people pare down the size of their air systems by trading those compressors and hoses for an onboard 20-volt battery and brushless motor. It holds up to 49 plastic-sleeved nails at a time, with a reported 899 nail-drives before needing to be recharged. It offers single trigger and multiple-fire bump modes, along with the ability to dial in nail length depending on your project needs. And you can throw it in your tool bag for anytime use.

The portability comes with a price, however. At 8.5 pounds you get no weight savings. Some have also noted that it still lacks the sheer driving force of an air-powered unit (although that wasn't the case for us). You can also get complete pneumatic setups with multiple tools for considerably less than this one unit, but if you're already in the DeWalt cordless ecosystem and have a large job looming, the electric advantages may be worth it for you. —*Mike Senese*

MAKE IT, WEAR IT: Wearable Electronics for Makers, Crafters, and Cosplayers

BY SAHRYE COHEN AND HAL RODRIGUEZ

$30 ampedatelier.com/make-it-wear-it

Authors Sahrye Cohen and Hal Rodriguez run San Francisco Bay Area-based design house Amped Atelier and their creations have walked the runways at various MakeFashion Galas in Calgary, Canada and at numerous Maker Faires around the world. The projects in their book aren't just wearable, they're *fashionable*. From 3D printing embellishments directly onto fabric to designing an LED matrix clutch purse, you'll want to show off these pieces.

Instructions are clear, well-thought-out, and easy to grasp — for a sewing and electronics novice like myself, the book never seemed overwhelming. Throughout the book foundational concepts for both disciplines are repeated enough to be helpful for beginners without being a slog for more advanced users. Many of the projects are sewing heavy, but no-sew options are also included for more electronics-inclined readers.

There are also a variety of materials to use and techniques to experiment with, and no two projects use exactly the same techniques or materials, so it always feels like you're attempting something different.

Whether you're new to wearables or ready to flex your skills with some fashionable inspiration, *Make It, Wear It* has something for you. —*Craig Couden*

MAKING TIME
BY BOB CLAGETT

$10 iliketomakestuff.com/product/making-time

If you're an avid viewer of YouTube videos from makers, you've likely heard of Bob Clagett, and know that he likes to make stuff. His new book, *Making Time*, tells the story of his journey to becoming a full-time maker, but it also details his concerns, his struggles, and his successes along the way, presented in a manner to be helpful to anyone who is interested in setting off on their own path to becoming a professional maker.

There's a lot to pore over here, from business strategy, work/life balance, even exploring some of the emotional vulnerability necessary to be a successful creative person. If you're an avid listener to the Making It podcast, many of these stories will be pretty familiar, but it's still worthwhile to have them all in one place.

—*Tyler Winegarner*

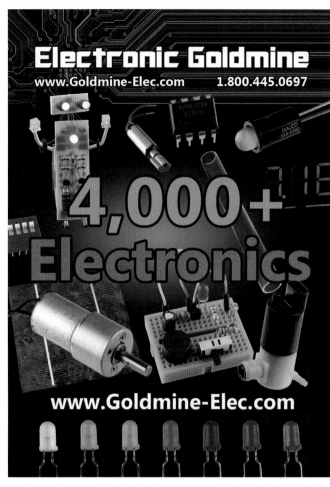

SHOW&TELL

Get inspired by some of our favorite submissions to Maker Share

If you'd like to see your project in a future issue of *Make:* magazine submit your work to makershare.com/missions/mission-make!

1 Prop builder **David Giles** decided to recreate the *Overwatch* character D.Va's Light Pistol for fun as a working Nerf blaster. Two flywheels propel the darts and a complex linkage system mimics the recoil of the barrel as you fire. The hardest part according to Giles was recreating the oddly shaped LED circuit patterns out of Veroboard. Overall, it's an impressive bit of engineering for small spaces. makershare.com/projects/dva-light-pistol-nerf-gun

2 **Geert Dom** loves tiny things. For years he's been making metal arms and armor for Lego and Playmobil sets. Far from toys, Dom hardens and tempers the steel himself. Accents like guards and pommels are crafted from brass while handles are made from wrapped wire and even carved wood. "Making tiny things makes you appreciate the finer things in life," he says, "like the error margin when drilling a 16" hole as opposed to drilling a 1/16" hole." makershare.com/projects/making-miniatures

3 Vintage metal lunch boxes have a certain decorative charm and make good storage space to boot, but **Beth Sallay** thought she could do better by adding custom draws and shelves. Sallay laser-cut the new additions out of 1/8" wood. Just remember which side is up when you go to open them. makershare.com/projects/lunchbox-upgrades

4 These light-up NeoPixel earrings from **Hilal Gungor** are easy to make and can be programmed with multiple colorful animations. makershare.com/projects/arduino-lilypad-controlled-neopixel-earrings-0

[+] Read about our Editors' Choice, Stowable Electronics Workshop, on page 72.

QUICK
BASE

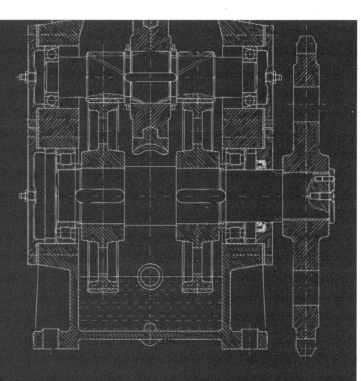

Build Your Own Time Machine

If you have a job, you probably spend too much time either in your inbox or a spreadsheet. Quick Base helps you get that time back by building your own powerful customized web applications with just the skills you already have - no coding required. *You'll save time, be a hero to your team, and get back to doing the work that matters most.*

Register for your free account
at quickbase.com/make

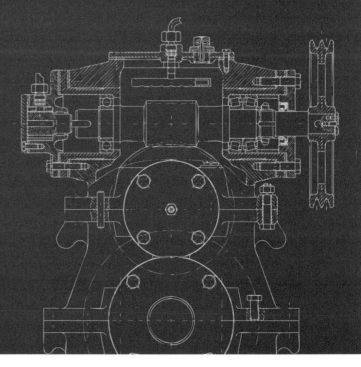

$9.99 US $11.99 CAN
ISBN: 978-1-68045-567-0